T0233301

Lecture Notes in Computer Science 10103

Commenced Publication in 1973
Founding and Former Series Editors:
Gerhard Goos, Juris Hartmanis, and Jan van Leeuwen

Editorial Board

David Hutchison
  Lancaster University, Lancaster, UK
Takeo Kanade
  Carnegie Mellon University, Pittsburgh, PA, USA
Josef Kittler
  University of Surrey, Guildford, UK
Jon M. Kleinberg
  Cornell University, Ithaca, NY, USA
Friedemann Mattern
  ETH Zurich, Zurich, Switzerland
John C. Mitchell
  Stanford University, Stanford, CA, USA
Moni Naor
  Weizmann Institute of Science, Rehovot, Israel
C. Pandu Rangan
  Indian Institute of Technology, Madras, India
Bernhard Steffen
  TU Dortmund University, Dortmund, Germany
Demetri Terzopoulos
  University of California, Los Angeles, CA, USA
Doug Tygar
  University of California, Berkeley, CA, USA
Gerhard Weikum
  Max Planck Institute for Informatics, Saarbrücken, Germany

More information about this series at http://www.springer.com/series/7407

Patrick Siarry · Lhassane Idoumghar
Julien Lepagnot (Eds.)

Swarm Intelligence Based Optimization

Second International Conference, ICSIBO 2016
Mulhouse, France, June 13–14, 2016
Revised Selected Papers

 Springer

Editors
Patrick Siarry
Université Paris-Est Créteil
Vitry-sur-Seine
France

Julien Lepagnot
Université de Haute-Alsace
Mulhouse
France

Lhassane Idoumghar
LMIA-INRIA Grand Est
Université de Haute-Alsace
Mulhouse
France

ISSN 0302-9743   ISSN 1611-3349 (electronic)
Lecture Notes in Computer Science
ISBN 978-3-319-50306-6   ISBN 978-3-319-50307-3 (eBook)
DOI 10.1007/978-3-319-50307-3

Library of Congress Control Number: 2016958515

LNCS Sublibrary: SL1 – Theoretical Computer Science and General Issues

© Springer International Publishing AG 2016
This work is subject to copyright. All rights are reserved by the Publisher, whether the whole or part of the material is concerned, specifically the rights of translation, reprinting, reuse of illustrations, recitation, broadcasting, reproduction on microfilms or in any other physical way, and transmission or information storage and retrieval, electronic adaptation, computer software, or by similar or dissimilar methodology now known or hereafter developed.
The use of general descriptive names, registered names, trademarks, service marks, etc. in this publication does not imply, even in the absence of a specific statement, that such names are exempt from the relevant protective laws and regulations and therefore free for general use.
The publisher, the authors and the editors are safe to assume that the advice and information in this book are believed to be true and accurate at the date of publication. Neither the publisher nor the authors or the editors give a warranty, express or implied, with respect to the material contained herein or for any errors or omissions that may have been made.

Printed on acid-free paper

This Springer imprint is published by Springer Nature
The registered company is Springer International Publishing AG
The registered company address is: Gewerbestrasse 11, 6330 Cham, Switzerland

Preface

These proceedings include a selection of the best papers presented at the International Conference on Swarm Intelligence Based Optimization, ICSIBO 2016, held in Mulhouse (France).

ICSIBO 2016 was a continuation of the conferences OEP 2003 (Paris), OEP 2007 (Paris), ICSI 2011 (Cergy-Pontoise), and ICSIBO 2014 (Mulhouse).

The aim of ICSIBO 2016 is to highlight the theoretical progress of swarm intelligence metaheuristics and their applications. Swarm intelligence is a computational intelligence technique involving the study of collective behavior in decentralized systems. Such systems are made up of a population of simple individuals interacting locally with one another and with their environment. Although there is generally no centralized control on the behavior of individuals, local interactions among individuals often cause a global pattern to emerge. Examples of such systems can be found in nature, including ant colonies, animal herding, bacteria foraging, bee swarms, and many more. However, swarm intelligence computation and algorithms are not necessarily nature-inspired.

Authors had been invited to present original work relevant to swarm intelligence, including, but not limited to: theoretical advances of swarm intelligence metaheuristics; combinatorial, discrete, binary, constrained, multi-objective, multi-modal, dynamic, noisy, and large-scale optimization; artificial immune systems, particle swarms, ant colony, bacterial foraging, artificial bees, fireflies algorithm; hybridization of algorithms; parallel/distributed computing, machine learning, data mining, data clustering, decision making and multi-agent systems based on swarm intelligence principles; adaptation and applications of swarm intelligence principles to real-world problems in various domains.

Each submitted paper was reviewed by three members of the international Program Committee. Two reviewing processes were undertaken: one before the conference and one after the conference.

We would like to express our sincere gratitude to our invited speakers: Brigitte Wolf and Maurice Clerc. The success of the conference resulted from the input of many people to whom we would like to express our appreciation: the members of Program Committee and the secondary reviewers for their careful reviews that ensure the quality of the selected papers and of the conference. We take this opportunity to thank the different partners whose financial and material support contributed to the organization of the conference: Université de Haute Alsace, Faculté des Sciences et Techniques et Institut Universitaire de Technologie de Mulhouse. Last but not least, we thank all the authors who submitted their research papers to the conference, and the authors of accepted papers who attended the conference to present their work. Thank you all.

August 2016   P. Siarry
                                                                                                      L. Idoumghar
                                                                                                      J. Lepagnot

Organization

Organizing Committee Chairs

P. Siarry
L. Idoumghar
J. Lepagnot

Program Chair

M. Clerc

Website/Proceedings/Administration

MAGE Team, LMIA Laboratory

Program Committee

Omar Abdelkafi	Université de Haute-Alsace, France
Ajith Abraham	Norwegian University of Science and Technology, Norway
Antônio Pádua Braga	Federal University of Minas Gerais, Brazil
Mathieu Brévilliers	Université de Haute-Alsace, France
Bülent Catay	Sabanci University, Istanbul, Turkey
Amitava Chatterjee	University of Jadavpur, Kolkata, India
Rachid Chelouah	EISTI, Cergy-Pontoise, France
Raymond Chiong	University of Newcastle, Australia
Maurice Clerc	Independent Consultant, France
Carlos A. Coello Coello	CINVESTAV-IPN, México
Jean-Charles Créput	Université de Technologie Belfort-Montbéliard, France
Rachid Ellaia	Mohammadia School of Engineering, Morocco
Frederic Guinand	Université du Havre, France
Jin-Kao Hao	Université d'Angers, France
Vincent Hilaire	Université de Technologie de Belfort-Montbéliard, France
Lhassane Idoumghar	Université de Haute-Alsace, France
Imed Kacem	Université de Lorraine, France
Jim Kennedy	Bureau of Labor Statistics, Washington, USA
Peter Korosec	University of Primorska, Koper, Slovenia
Abderafiaâ Koukam	Université de Technologie Belfort-Montbéliard, France
Nurul M. Abdul Latiff	Universiti Teknologi, Johor, Malaysia
Fabrice Lauri	Université de Technologie de Belfort-Montbéliard, France

Stephane Le Menec	RGNC at EADS/MBDA, France
Julien Lepagnot	Université de Haute-Alsace, France
Evelyne Lutton	INRA-AgroParisTech UMR GMPA, France
Vladimiro Miranda	University of Porto, Portugal
Nicolas Monmarché	Université François Rabelais Tours, France
René Natowicz	ESIEE, France
Ammar Oulamara	Université de Lorraine, France
Yifei Pu	Sichuan University, China
Maher Rebai	Université de Haute-Alsace, France
Said Salhi	University of Kent, UK
René Schott	University of Lorraine, France
Patrick Siarry	Université de Paris-Est Créteil, France
Ponnuthurai N. Suganthan	Science and Technology University, Singapore
Eric Taillard	University of Applied Sciences of Western Switzerland
El Ghazali Talbi	Polytech'Lille, Université de Lille 1, France
Antonios Tsourdos	Defence Academy of the United Kingdom, UK
Mohamed Wakrim	University of Ibou Zohr, Agadir, Morocco
Rolf Wanka	University of Erlangen-Nuremberg, Germany

Contents

Plenary Talks

Total Memory Optimiser: Proof of Concept and Compromises

Maurice Clerc[✉]

Independent Consultant, Groisy, France
Maurice.Clerc@WriteMe.com

Abstract. For most usual optimisation problems, the Nearer is Better assumption is true (in probability). Classical iterative algorithms take this property into account, either explicitly or implicitly, by forgetting some information collected during the process, assuming it is not useful any more. However, when the property is not globally true, i.e. for deceptive problems, it may be necessary to keep all the sampled points and their values, and to exploit this increasing amount of information. Such a basic Total Memory Optimiser is presented here. We experimentally show that this technique can outperform classical methods on small deceptive problems. As it gets very computing time expensive when the dimension of the problem increases, a few compromises are suggested to speed it up.

1 Motivations

As of today (2016-08), all iterative optimisers do forget some positions they have previously sampled in the search space, sooner or later. This is true for even a method like Tabu Search [6]. This is a loss of information about the "shape" of the landscape of the problem at hand. An obvious drawback is that the same position may be sampled and evaluated several times, which is useless. In order to prevent this undesirable behaviour, classical algorithms progressively define "bad areas" in which the probability to be sampled is null or extremely small.

On the one hand, this may introduce a risk of being wrong, but, on the other hand, there is no doubt that most of these methods are efficient in practice on many problems. We claim here that it is because for these problems a "Nearer is Better in probability" assumption is valid.

But what if it is not true? In such a case, it may be useful to make use of an algorithm that takes into account all the information that is collected during the iterative search.

2 Nearer is Better Assumption

The NisB assumption has been studied in [2,3]. Let us just give here a short definition and a few examples.

© Springer International Publishing AG 2016
P. Siarry et al. (Eds.): ICSIBO 2016, LNCS 10103, pp. 3–19, 2016.
DOI: 10.1007/978-3-319-50307-3_1

2.1 Definition

What we call here the NisB correlation can be estimated as follows:

- sample at random (uniform distribution) N times three positions and sort them so that $f(x_1) \leq f(x_2) \leq f(x_3)$;
- let n_{isB} be the number of times we have $f(x_1) \leq f(x_2) < f(x_3)$ and $distance(x_1, x_2) < distance(x_1, x_3)$.
- then the correlation is defined so that it is in $[-1, 1]$ by

$$\rho = -1 + 2\frac{n_{isB}}{N} \tag{1}$$

2.2 Examples

For a strictly monotonic function, the NisB correlation is of course equal to 1. For most usual problems it is positive. However, as we can see from Table 1 it can easily be negative, as soon as there are some plateaus in the function[1]. The precise definitions of these four problems are in the Appendix A.1.

Also, if it is not equal to 1, that means it is sometimes locally negative. For example, for the Parabola in the Fig. 1 the global correlation is $17/18 = 0.944$, not equal to 1 because of triplets of points "around" the minimum.

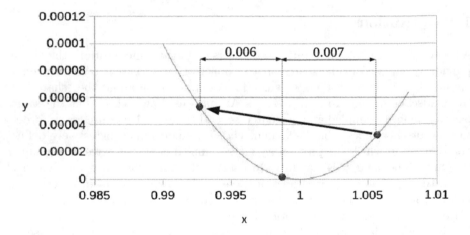

Fig. 1. When Nearer is Worse, locally. However, globally, the correlation is positive.

Actually, it may be an interesting exercise to evaluate the correlation as a function of the vicinity fraction, where we define this vicinity fraction as a proportion of the search space around each point. So, the global correlation

---

[1] This is an open question: is it possible to define a Lipschitzian function without any plateau but with a negative NisB correlation?.

Table 1. Examples of Nearer is Better correlation.

Name	Representation	ρ
	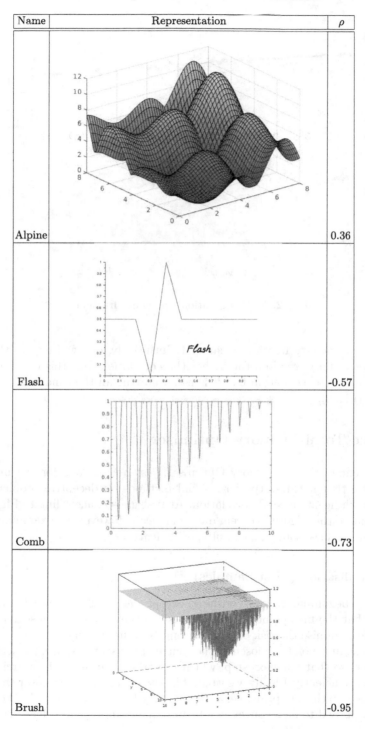	
Alpine		0.36
Flash		-0.57
Comb		-0.73
Brush		-0.95

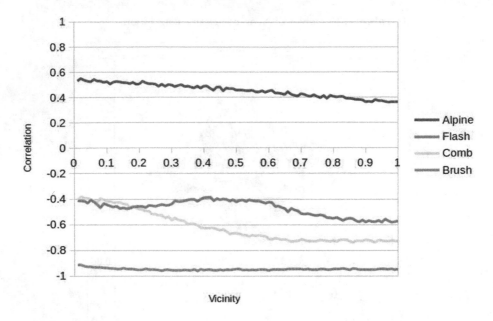

Fig. 2. NisB correlation vs vicinity fraction.

is given for a vicinity fraction equal to 1. Intuitively, one will expect that the smaller the vicinity fraction, the higher the correlation; but this does not hold in a very strong sense, and in fact it is not even true all the time (see the Flash example) (Fig. 2).

3   Basic Total Memory Optimiser

We now define a Total Memory Optimiser, just a basic one for the moment, in the hope that it can outperform usual methods on deceptive problems, i.e. the ones with negative NisB correlation. At first glance, according to Fig. 3, the components of this algorithm seem quite classical. However, by examining them in detail we will see some important specific features.

3.1   Initialisation (First Samples)

At the very beginning we know nothing, except the search space. It is then easy to prove that the first point to sample must be the centre of this search space, in order to minimise the risk of being wrong (see the Appendix A.2).

After the first point is chosen at the centre of the search space, we need some more points so that our next step, which is a triangulation of the search space, can be done efficiently. Ideally, we would like the triangulation to cover the whole search space, so that, in principle, any point may be then sampled. As the search space is supposed to be a convex polyhedron, a certain way to achieve that is to

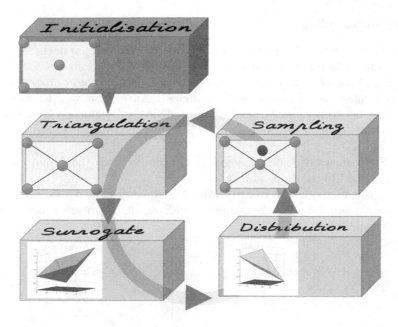

Fig. 3. Basic total memory optimiser.

sample all the corners. In practice the search space is a D-rectangle (which can easily be transformed into a D-square, in order to later simplify strategies like "search around"). So, finally, the number of initial points is $2^D + 1$.

Of course, it increases very rapidly with D, and it may be unacceptable for small computers. But for the moment, to describe this basic TMO, we do not care. Later, we will look at some possible compromises (Fig. 4).

Fig. 4. Initialisation: centre + corners.

3.2 From Sampling to Triangulation

At a given time step we have sampled N points. We will explain in the next step that the surrogate function is made of "triangular" facets (interval for $D = 1$, real triangles for $D = 2$, tetrahedron for $D = 3$, etc.), whose projections on the search space are of course also "triangles". An elegant way to ensure no overlapping between these "triangles" is to define a Delaunay's triangulation between these N points [4]. Actually, this is a simplified variant of the approach described in [1].

The Fig. 5 shows two such triangulations, coming from the Alpine problem that is described in the Appendix A.1.

Again, this method is very costly when the dimension increases, but as said before, we do not care for the moment, and, again, there are some possible cheaper compromises.

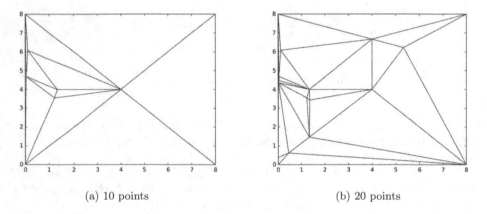

(a) 10 points   (b) 20 points

Fig. 5. Delaunay's triangulations (Alpine problem, solution point on $(1, 2)$).

3.3 From Triangulation to Surrogate Function

Using surrogate functions is not at all new. See for example [7]. However, here, we modify it after each new sampling and, moreover, we want to use the best one under as few hypotheses as possible. Indeed, most "aesthetic" surrogate functions do suppose some superfluous properties like differentiability everywhere, at least implicitly.

To simplify the reasoning, let us consider a 1D problem whose value space is $[y_{min}, y_{max}]$, two sampled points x_1, x_2, and their values $f(x_1)$, $f(x_2)$. We do not know the values of f between the two points, and we want to estimate them by using the surrogate function g that minimises the risk of being wrong, i.e. the risk of sampling x with $g(x) \neq f(x)$. We must have $g(x_1) = f(x_1)$ and $g(x_2) = f(x_2)$, but for any other point x in $]x_1, x_2[$ we just know that we must have $y_{min} \leq g(x) \leq y_{max}$. If we do not make any assumption, it is easy

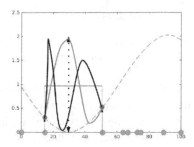

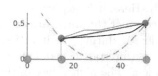

(a) No hypothesis. The best surrogate function is a plateau.

(b) NisB correlation locally positive, and minimal variability.

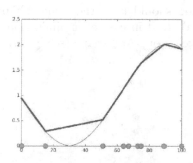

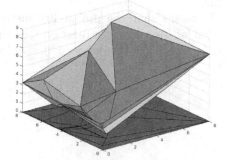

(c) Piece-wise linear surrogate function (1D).

(d) Piece-wise linear surrogate function (2D).

Fig. 6. Choosing the "best" surrogate function. Without any hypothesis, it is a plateau. With the locally positive NisB correlation hypothesis, it is monotonic. And with the minimum variability constraint, it is a piece-wise linear one.

to prove that the best g is given by $g(x) = \frac{y_{min}+y_{max}}{2}$ (see the Fig. 6a and the Appendix A.2). This is not satisfying for this plateau function does not increase the quantity of information (even if only guessed) we can use.

So, we do make a hypothesis. Here, it is that the NisB correlation between the two points is locally positive[2]. It implies that g must be monotonic. But there is still an infinity of possible functions (Fig. 6b). So, we also apply the Occam's razor, or, more formally, we assume that the *variability* (see the Appendix A.3 for a formal definition) of the function must be minimal. Then, the only possible function is the linear one. More generally, on D dimensional problems, the surrogate function is made of "triangular" facets (see Figs. 6c and d).

---

[2] It may seems contradictory with the fact that we want to cope with problems for which the NisB correlation is globally negative. Even in such a case, it is sometimes *locally* positive, and more and more when the number of points increases.

3.4   From Surrogate Function to Estimation of Distribution

This step is of course crucial. TMO is stochastic, and therefore it has to make use of an estimation of distribution, as all stochastic methods do, either implicitly or explicitly. Here, it is explicit.

Let us consider the surrogate function of the Fig. 6c. It contains all the information collected during the iteration: the sure ones (sampled points), and the probable ones (interpolations). Let us first suppose that the problem is a maximisation problem. The high values of the surrogate function tell us where probably the high values of the true function lie. So, it seems reasonable (still because of the local NisB principle) to sample "around" the points of high value with high probability, and conversely with less probability "around" the points of low value.

It means that the probability distribution should have the same "shape" as that of the surrogate function (see Fig. 7a). And in case of minimisation, it should be exactly the opposite: it should have an opposite shape (see Fig. 7b).

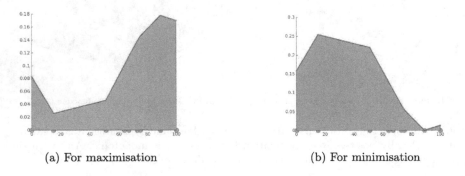

(a) For maximisation   (b) For minimisation

Fig. 7. Estimations of distribution, derived from the surrogate function of the Fig. 6c.

So, we have a straightforward way to define the estimation of distribution: just transform the surrogate function into a probability distribution, either directly, or by taking its opposite (in practice by applying a formula like $minimum + maximum - g$).

Note that it is perfectly possible to "distort" the distribution thanks to a user-defined morphing parameter, which can be either optimistic or pessimistic. To keep the presentation simple, we will not do that in the examples.

3.5   From Estimation of Distribution to Sampling

This step is a classical one. Moreover, as the distribution is entirely defined by the sampled points, it is technically easy. In order to improve the efficiency of TMO, we also make use of "representatives" of the triangular facets. A representative can for example be:

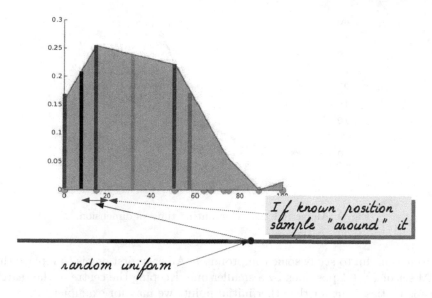

Fig. 8. Sampling from the estimation of distribution.

1. the centre of the facet;
2. a weighted gravity centre (weights depending on the values of the vertices);
3. a random point inside the facet.

In what follows, we simply use method 1 (see the Fig. 8). We virtually put all the values (of the sure points and of the representatives) along a line, side by side, draw at random a number (uniform distribution) between 0 and their sum, and go back to the corresponding value.

If it is a representative, we indeed sample this new point. If it is an already known point x_1, we sample "around" it. In practice, the search space has been normalised as a D-square (if $D \geq 2$), and "around" means "inside a small D-square centred on the known point". If $D = 1$, we of course just consider intervals.

In the examples, "small" is defined as follows:

– find the nearest known point x_2;
– the edge of the D-square is $2\alpha \|x_1 - x_2\|$, where α is a user-defined parameter. In the examples of this paper $\alpha = \frac{1}{3}$.

Note that instead of using the $g(x)$ values it is perfectly possible to sort them and to use their ranks. Actually some experiments suggest that it may be a better way, and this is what is done here.

4   Compromises

As soon as the dimension of the problem increases, it is clear that the computing cost becomes rapidly unacceptable in practice, as we can see on the Fig. 9.

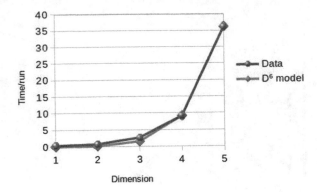

Fig. 9. Alpine function. Computing time vs dimension.

So it is useful to study some compromises. An easy first one is to replace the initial set of $2^D + 1$ positions by a smaller one. Keeping the centre of the search space is costless, but for the other initial points we may for example:

- sample at random N points (a very usual method, in fact);
- sample at random N points and C corners (typically $C = 1$).

It means that the "covering" of the search space by triangulation is now just a partial one, but if we launch several runs, we hope that the solution will be inside a triangle, and has a non null probability to be found. Actually, it is the way classical optimisers do work.

A complete re-computation of Delaunay's triangulation after each new sampled point is introduced is a costly process. Instead, we can replace it by a local re-triangulation. A very simple one is shown in the Fig. 10. We just subdivide the "triangle" in which the new point lies.

5   Comparisons

We can now make some comparisons between TMO (basic and some variants with compromises) and a few classical optimisers. For fair comparisons, we have to carefully specify two things:

- the budget;
- the user's demand.

5.1   Budget

A budget for iterative optimisation on a computer has at least three components:

- a maximum number of evaluations E;
- a maximum computing time T;
- a maximum memory size M.

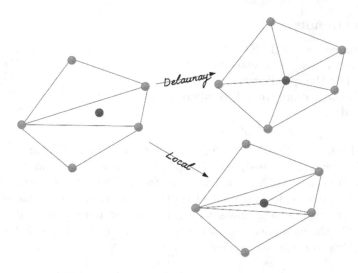

Fig. 10. Simplifying the re-triangulation.

In this paper, we want to check how efficient TMO is in taking into account all the information collected during the iterative process. So, for comparisons, we will just use the same E for several methods. Actually, this is a classical approach.

5.2 Demand

Many studies take the mean result over n runs or the median value as the comparison criterion, and perform statistical analyses to estimate to which extent such values are reliable. But in practice a user does not care about means or medians. His/her question is (in case of minimisation): "With *this* budget and *this* algorithm, what is the probability of getting a result smaller than ε"?

It is out of the scope of this paper to analyse in detail how to spend a given budget. We just apply the classical method here: n independent runs of E/n evaluations. But keep in mind this is far from being the best way always.

To answer the user's question, we need to estimate the probability distribution of the results. In practice, it means we need to build an estimation of the PDF (Probability distribution function) or, easier to use, an estimation of the CDF (Cumulative Distribution Function). By doing that, when comparing two algorithms A_1 and A_2, it often happens that the CDF curves have at least one cross point, for a given result ε^*. Let us suppose $\mathrm{CDF}(A_1)$ is "above" $\mathrm{CDF}(A_2)$ for results smaller than ε^*, and the contrary for higher values. If the user is very demanding (i.e. does accept only values smaller than $\varepsilon_{max} \leq \varepsilon^*$), then A_1 is the best choice. But if the user accepts values smaller than $\varepsilon_{max} > \varepsilon^*$, then, on the contrary, A_2 is the best choice.

5.3 Some Results

We first compare a few TMO variants, with some compromises, and the basic TMO. As we will see, some variants may be better than this basic TMO. But to better highlight the specificity of TMO, i.e. its efficiency on problems of negative global NisB correlation, we will compare then three classical methods only to the basic TMO.

TMO variants. On the Fig. 12, "init 2" is for the initialisation method of the basic TMO (centre + corners), and "init 0" is for centre + random initialisation (here with 89 positions). And "reTriang k" means Delaunay's re-triangulation every k new positions, local simplified re-triangulation otherwise. In particular "reTriang *Inf*" means Delaunay's triangulation just once, after initialisation, and then only the simplified method is used. The Fig. 11 shows that then the computing time is easily far smaller.

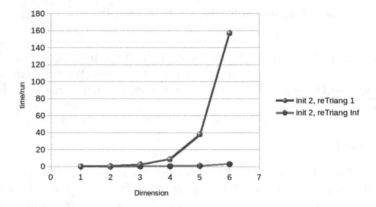

Fig. 11. Using simplified retriangulation dramatically reduces the computing time (Alpine, 150 evaluations).

Moreover, at least for problem with a positive NisB correlation like Alpine, not only complete and systematic re-triangulation is not necessary, but partial ones may give better results: higher probability for small values, as we can see on the Fig. 12.

When the NisB correlation is negative, as for the Brush function, we have something similar. On the Fig. 13 the three curves with "init 0" are better than the ones with "init 2". Note that, though, we are here in a case where the number of corners is far smaller than the number of random initial points. The conclusion may not be valid for dimension greater than 6. Unfortunately, the laptop used for this study can not cope with Delaunay's triangulations in dimension 7 or greater.

So, we now consider only the basic TMO for comparisons with classical methods, i.e. "init 2, reTriang 1".

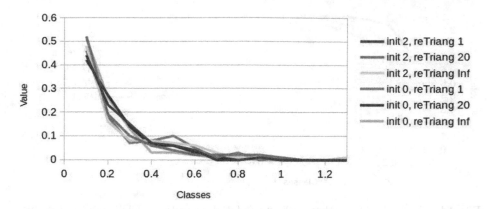

Fig. 12. Alpine, TMO variants, 100 runs of 150 evaluations, PDFs curves.

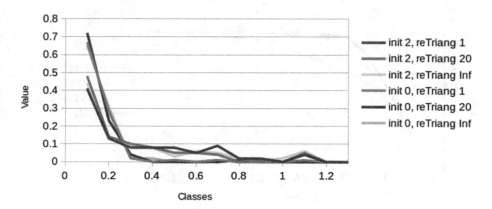

Fig. 13. Brush, TMO variants, 100 runs of 150 evaluations, PDFs curves.

TMO vs Others. The Fig. 14 represents the CDFs of four methods on the Alpine function. As said, the first one is the basic TMO, and the other ones are GA-MPC[3], CMA-ES [8], and APS 11. GA-MPC [5] was the winner of the CEC 2011 competition, CMA-ES was the winner of the CEC 2005 competition[4]. For APS [9], a well-documented MATLAB© implementation can be downloaded from http://aps-optim.info. It easily outperforms both GA-MPC and CMA-ES on the CEC 2011 problems, but the point here is that TMO is clearly the worst method on Alpine, unless results greater than 0.4 are acceptable: in that case it outperforms CMA-ES (but still not the other ones).

---

[3] We would like to thank Dr. Saber Elsayed for providing the MATLAB© code of GA-MPC.

[4] In fact, we used a more recent and better version 3.62, downloaded from https://www.lri.fr/~hansen/cmaes_inmatlab.html.

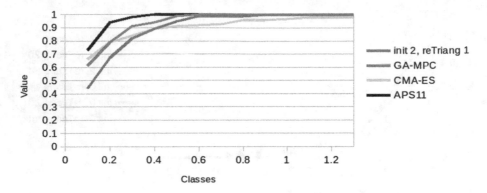

Fig. 14. Alpine, TMO vs others, 100 runs of 150 evaluations, CDFs curves. On this "positive" NisB problem, TMO is the worst method, unless the user is not very demanding, accepting results greater than 0.4. In that case, CMA-ES is the worst.

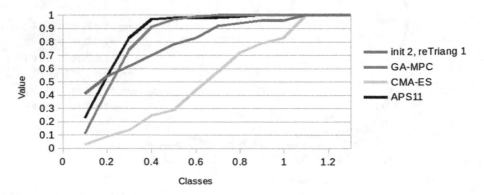

Fig. 15. Brush, TMO vs others, 100 runs of 150 evaluations, CDFs curves. On this "negative" NisB problem, TMO is the best method to find good results (smaller than 0.2), but for higher acceptable results, APS11 and even GA-MPC are better. For any demand, CMA-ES is largely the worst.

The conclusions are different on a very deceptive function like Brush (see the Fig. 15). Now TMO is the best choice as soon as we accept only good (small) results. Note that on such a problem CMA-ES is particularly bad.

When and How to Use TMO?

In this study we have given a precise definition of "deceptive problems", by saying that these are the ones for which the global NisB correlation is negative. However, in the field of optimisation, "deceptive problem" usually just implies that classical algorithms perform worse than random search on it, or even simply "very difficult" [10]. In particular, combinatorial problems are often said to be deceptive.

Preliminary tests show that for many combinatorial problems, if not all, the NisB correlation is positive. Of course, this is done by using a specific distance measure, for example the Kendall tau one's, or, in a more intuitive way, the minimum number of transpositions to transform a given "position" (typically a permutation of K elements) into another one.

On the one hand, we have said that TMO is for problems with negative NisB correlation, so it is probably not a good tool for combinatorial problems. On the other hand, for these problems, as the correlation is just *slightly* positive it may be nevertheless worth trying to apply it.

Also, TMO could be combined with a classical iterative algorithm A, by using the following strategy:

– compute the NisB correlation after each new evaluation;
– if positive, use A;
– if negative, use TMO.

Again, nothing proves it will work, but, again, it seems worth trying.

If it appears that TMO, either alone or in combination, indeed gives good results on some real world problems, it will be necessary to increase its computational efficiency. We already have seen a few possible ways, but more are possible, and should be tested.

A   Appendix

A.1   Problem Definitions

Alpine. For dimension D, the search space is $[0, 4D]^D$. Function f is defined as:

$$f(x_1, \ldots, x_D) = \sum_{d=1}^{D} |x_{d,\delta} \sin(x_{d,\delta})| + 0.1 |x_{d,\delta}| \qquad (2)$$

with $x_{d,\delta} = x_d - \delta d$. In this case, we have simply chosen $\delta = 1$. This parameter serves to ensure that the minimum is not at the centre of the search space or on a diagonal. The problem is multimodal and non-separable.

Deceptive 1 (Flash). The search space is $[0, 1]$. Function f is defined as:

$$\begin{cases} x \leq 2c_1 & \rightarrow f(x) = c_2 \\ 2c1 < x \leq 3c_1 & \rightarrow f(x) = c_2 - \frac{c_2}{c_1}(x - 2c_1) \\ 3c_1 < x \leq 4c_1 & \rightarrow f(x) = \frac{2c_2}{c_1}(x - 3c_1) \\ 4c_1 < x \leq 5c_1 & \rightarrow f(x) = 2c_2 - \frac{c_2}{c_1}(x - 4c_1) \\ x \geq 5c_1 & \rightarrow f(x) = c_2 \end{cases} \qquad (3)$$

with, in this case, $c_1 = 0.1$ and $c_2 = 0.5$. The problem is unimodal, but with plateaus.

Deceptive 2 (Comb). The search space is $[0, 10]$. Function f is defined as:

$$f(x) = \min\left(c_2, 1 + \sin(c_1 x) + \frac{x}{c_1}\right) \tag{4}$$

with, in this case, $c_1 = 10$ and $c_2 = 1$. The problem is multimodal, but with plateaus.

Deceptive 3 (Brush). The search space is $[0, 10]^2$. Function f is defined as:

$$f(x_1, x_2) = \min\left(c_2, \sum_{d=1}^{2} |x_d \sin(x_d)| + \frac{x_d}{c1}\right) \tag{5}$$

with, in this case, $c_1 = 10$ and $c_2 = 1$. The problem is multimodal and non-separable.

A.2   When we Know Nothing, the Middle is the Best Choice

On the Search Space. Let x^* be the solution point (we do suppose here it is unique). If we sample x, the error is $\|x - x^*\|$. At the very beginning, as we know nothing, the probability distribution of x^* is uniform on the search space. Roughly speaking, it can be anywhere with the same probability. So, we have the sample x in order to minimise the risk given by

$$r = \int_{x^* \in S} \|x - x^*\| \tag{6}$$

Let us solve it for $D = 1$, and $S = [x_{min}, x_{max}]$. We have

$$r = \int_{u=x_{min}}^{x} (x - u)\, du + \int_{u=x}^{x_{max}} (u - x)\, du$$
$$= \left[xu - \frac{u^2}{2}\right]_{u=x_{min}}^{x} + \left[\frac{u^2}{2} - xu\right]_{u=x}^{x_{max}}$$
$$= x^2 - (x_{max} + x_{min})\,x + \frac{x_{max}^2 + x_{min}^2}{2}$$

And the minimum of this parabola is given by

$$x = \frac{x_{max} + x_{min}}{2}$$

For $D > 1$ the proof is technically more complicated (a possible way is to use recurrence and projections), but the result is the same: the less risky first point is the centre of the search space.

On the Value Space. The same reasoning can be applied to the value space, when we do not make any hypothesis like say a positive local NisB correlation, and when we know the lower and upper bounds of the values, respectively y_{low} and y_{up}. On any unknown position of the search space the distribution of the possible values on $[y_{low}, y_{up}]$ is uniform and therefore the less risky is, again, the middle, i.e. $\frac{y_{low} + y_{up}}{2}$.

A.3 Variability of a Landscape

We use here a specific definition, which is different from the definition of variance in probability theory. Let f be a numerical function on the search space S. What we call *variability* on a subspace s of S is the quantity

$$v = \int_{s^4} \left| \frac{f(x_2) - f(x_1)}{\|x_2 - x_1\|} - \frac{f(x_3) - f(x_1)}{\|x_3 - x_1\|} \right| \tag{7}$$

where $\{x_1, x_2, x_3\}$ is an element of $s^3 = s \otimes s \otimes s$ (Euclidean product), under the constraint $x_3 = x_1 + \lambda (x_2 - x_1)$ or, equivalently, $(x_2 - x_1) \times (x_3 - x_2) = 0$ (cross product). The definition may seem to be complicated, but it just means that in any direction the slope of the landscape is constantly the same.

References

1. Beyhaghi, P., Cavaglieri, D., Bewley, T.: Delaunay-based derivative-free optimization via global surrogates, part I: linear constraints. J. Glob. Optim., 1–52 (2015)
2. Clerc, M.: When Nearer is Better, p. 19 (2007). https://hal.archives-ouvertes.fr/hal-00137320
3. Clerc, M.: Guided Randomness in Optimization. ISTE (International Scientific and Technical Encyclopedia). Wiley (2015)
4. de Berg, M., Cheong, O., van Kreveld, M., Overmars, M.: Computational Geometry. Springer, Heidelberg (2008)
5. Elsayed, S.M., Sarker, R.A., Essam, D.L.: GA with a New Multi-Parent Crossover for Solving IEEE-CEC2011 Competition Problems (2011)
6. Glover, F., Laguna, M.: Tabu Search. Kluwer Academic Publishers (1997)
7. Han, Z.-H., Zhang, K.-S.: Surrogate-based optimization. INTECH Open Access Publisher (2012)
8. Hansen, N.: The CMA Evolution Strategy: A Tutorial. Technical report (2009)
9. Omran, M.G.H., Clerc, M.: An adaptive population-based simplex method for continuous optimization. Int. J. Swarm Intell. Res. **7**(4), 22–49 (2016)
10. Weise, T., Zapf, M., Chiong, R., Nebro, A.J.: Why is optimization difficult? In: Kacprzyk, J., Chiong, R. (eds.) Nature-Inspired Algorithms for Optimisation. SCI, vol. 193, pp. 1–50. Springer, Heidelberg (2009)

Inspiration by Swarms

Brigitte Wolf[(✉)]

University of Wuppertal, Wuppertal, Germany
bwolf@uni-wuppertal.de

For me swarms and most of all swarms of fishes are a fascinating phenomenon.

I am a designer and my area of research is design theory and most of all strategic design. The task of strategic design is to implement design as the corporate strategy.

During the last years the awareness for design has increased. More and more companies have realized that they can benefit a lot from design. At the same time the complexity of design has increased. To make best use of design it needs to be implemented strategically into the business strategy. Statistics have proved that design driven companies, which use design in a strategic way are more successful on the long run (Figs. 1 and 2).

Companies are unique and make different use of design. The Denish Design Center has created the design ladder to categorize the companies the way the use design.

Quite a number of companies have arrived at level four and are unsure how to proceed. The great challenge is to move to the next level. The big question is: How to do that? Unfortunately there is no recipe and no textbook for advice. The facts we have to take into account:

- our resources and energy sources are limited;
- the development of new technologies is proceeding fast;
- the use of new technologies changes consumption patterns equally fast.

It is in our hands to shape the future in way that our children and grandchildren can enjoy their lives in dignity and peace. The old methods and strategies are useless to design the future. The challenge is to consider the contradictory demands regarding available resources, customer demands, technological development, changes in life-style, politics and economy. New strategies are needed to create solutions, which balance the different demands (Fig. 3).

My provocative hypothesis is: Companies need to use design as swarm intelligence to climb from level four to level five!

That is not an easy task. Changes demand exchange! That is why the swarms come into play. The great challenge for companies is to deal with facts and figures, and with information and decisions in highly complex environments under changing conditions. That is the everyday life of schools of fish. For schools of fish that seems not be a problem at all.

In my holidays I like to dive in clear blue and warm waters with great visibility. For me it is extremely fascinating to observe the schools of fish. How they move, how they interact, how they react and how they organize themselves. The movement of schools of fish is pure joy for designers' eyes. The way they move is applied aesthetics. I want to share my experiences with you for a moment.

© Springer International Publishing AG 2016
P. Siarry et al. (Eds.): ICSIBO 2016, LNCS 10103, pp. 20–38, 2016.
DOI: 10.1007/978-3-319-50307-3_2

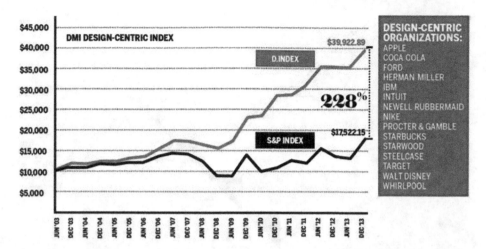

Fig. 1. Economic development of companies with a high design index compared to S&P (dmi journal)

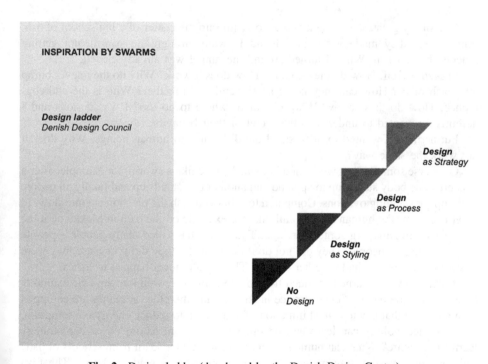

Fig. 2. Design ladder (developed by the Denish Design Center)

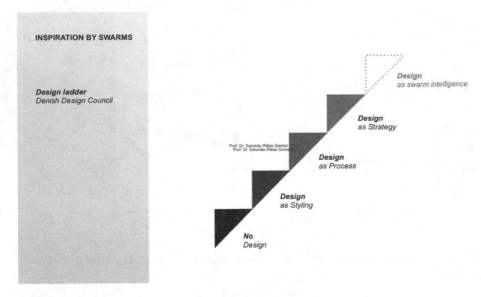

INSPIRATION BY SWARMS

Design ladder
Denish Design Council

Design
as swarm intelligence

Design
as Strategy

Prof. Dr. Salomão Ribas Gomez
Prof. Dr. Salomão Ribas Gomez

Design
as Process

Design
as Styling

No
Design

Fig. 3. Design ladder – future development

One of my greatest experiences was to swim into the center of a big school of fish. Immediately they made a big tunnel and I swam through the swarm, not getting touched by any fish. When I turned around the tunnel was already closed.

I asked myself: How do they do that? How does it work? Why do they never bump into each other? How can they move so elegantly all together? Who is the choreographer? How do they know? Who tells them where to go too? I was curious and I definitively wanted to understand the secret of their behavior.

Furthermore I wanted to understand the difference to human beings. Why does it not work the same way?

Of course sometimes it works and humans behave like a swarm, for example: after a concert some body stands up to applaud and another one stands up and finally all people stand up for standing ovations. Compared to schools of fish the reaction is quite slow. In other occasions the human swarms fail, like for example on the hadj in Mekka or at the love-parade in Duisburg some years ago. The crowd has killed many people – people were stepping on people. They did not do it deliberately, they did it because they were pushed by others and had no other chance. That would never happen to fishes.

Technology has learned so much from nature. Bionic is well known. One example: Shark skin is the most efficient surface in streams and therefore airplanes are equipped with a surface that is a technical imitation of shark skin to reduce energy consumption.

When technology can learn and benefit so much from nature. What can strategy learn from nature? What can human organization structures learn from schools of fish? This question was causing my interest even more when I read the book "The Fish Inside You", written by the palaeontologist Neil Shubin [1]. He figured out that we all stem from fishes. Millions of years ago some fishes started to leave the water to discover Earth and they decided to live on Earth. They were already bearing inside all beginnings of physical body elements, which characterize the diverse beings and creatures on Earth that have developed throughout the evolution process, like animals, birds and human beings.

I wanted to understand, how the fishes manage their being together. In the following I want to summarize, what I have learned from the swarms and I would like to point out the difference to human beings. Then I would like to give some examples where swarm intelligence is already in use and finally I will talk about my first ideas how I used swarm behavior as source of inspiration for a real case. The example is a company that intends to jump from level four to level five on the design ladder. I will work on this case again with our master students in the winter term 2016/17.

1 Swarm Behavior

Swarms exist since millions of years. Compared to swarms the life of companies is rather short. Swarms are self-organizing systems. They have no leader and no master-plan. This system seems to be very sustainable. I know you are all familiar with the rules of swarm behavior. Please excuse my repetition of basic knowledge, but I use it to point out the difference to human beings.

Craig Reynolds discovered that fishes follow only a few rules to be able to act as a swarm [2]:

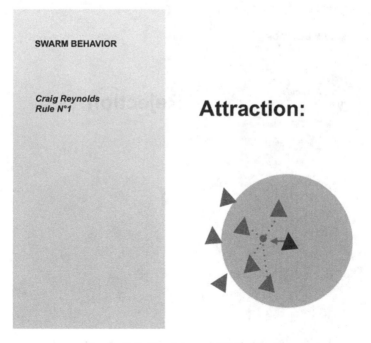

Fig. 4. Rule No. 1 'stay with the others'

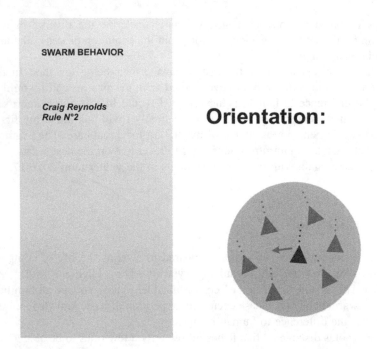

Fig. 5. Rule No. 2 'swim to the average of the direction of the others'

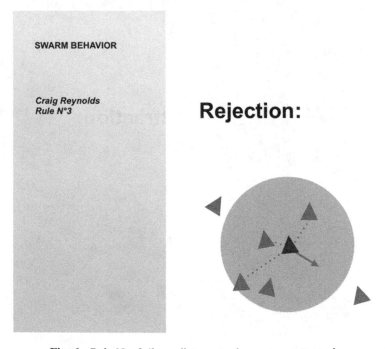

Fig. 6. Rule No. 3 'keep distance to the ones next to you'

Reynolds used this three rules to instruct his "boids" - boids symbolize fish in computer simulations. He also assumed that each fish has a rejection zone around and that the fish reacts as soon as someone enters this zone (Figs. 4, 5, 6 and 7).

The result of his experiment was, that the boids stayed together as a swarm. They were even able to evade obstacles and to stick together. (These insights are used to plan emergency escape routes for stadions and event halls.)

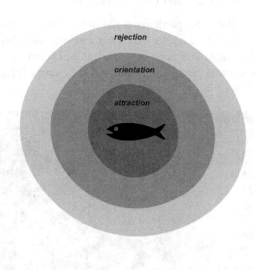

Fig. 7. Rejection zone

Jens Krause [3] confirmed these findings by stating that the rejection zone consists of different layers:

- attraction;
- orientation;
- rejection.

A fish is in the center of the rejection zone.

Iain Couzin [4] went a step further. He wanted to investigate the role of the distance between the fishes. Couzin varied the rejection zones in computer simulations.

The behavior of the swarm changed immediately when he reduced or enlarged the rejection zone. The variation of the rejection zone determined the transition from one formation to another. Transition happens because the swarm decides to search for food, to escape from an enemy, to hang around or to move to another place.

Schools of fish have obviously a very efficient way of communication. Fish swarms with thousands of fishes turn around in the same moment due to perfect communication system. It is a miracle to observe these movements. I conclude: Communication in swarms of fish is very efficient – communication in swarms of people is rather poor.

Jens Krause wanted to find out, if people in a crowd move the same way as fishes or if they behave differently. He started an experiment on the fair ground in Cologne. Two hundred people participated voluntarily [5].

Two hundred people were instructed with rules similar to swarms:

- move with normal speed and stick to the group;
- keep the distance to persons next to you considering the length of an arm;
- don't talk and don't gesticulate.

When the participants did not receive any further instruction the participants started to walk in a circle – an inner circle in one direction and an outer circle in the other direction. The same behavior can be observed with barracudas. They swim in an inner and outer circle and sometimes they cross (Fig. 8).

Fig. 8. Swarm Experiment in the Cologne Fair organized by Jens Krause and reported in the TV program Quarks&Co by Ranga Yogeshwar

In the next step of the experiment three people were instructed to move to a certain destination. Some followed but the crowd remained together.

In the following 10 people – 5% of the group – got the instruction to move to a certain destination and the whole swarm of people followed them and met at that place.

In further research projects Jens Krause figured out that even in groups bigger than 200 beings10 individuals are sufficient to guide the swarm to a certain destination [5].

Human beings can behave like swarms if they accept to follow similar rules.

But human beings are different from fish. Fishes depend on the well being of the swarm. Human beings are selfish, they are primarily interested in their own well-being.

The experiment of Jens Krause demonstrated further that the independence of decision is important for swarm behavior. If the participants would have had the chance to talk to each other, they probably would have behaved differently.

A comfort zone surrounds equal to fishes human beings. As long as the comfort zone is not disturbed human beings feel ok in a swarm. When the comfort zone is attacked, humans react different then swarms. Human beings are not able to communicate efficiently in huge groups. They panic and provoke terrible accidents like for example in the hadj.

Schools of fish communicate obviously very efficiently. In the ocean environment a lot of information is just white noise. It is easy to fail and to detect danger where no danger is. One single fish can fail easily, but the probability that many fishes make the same mistake is rather unlikely. This phenomenon is well known as the wisdom of the crowd [6]. That means in case of an enemy attack, a single fish would not particularly take a good decision, but the bigger the swarm, the bigger the probability that all fishes together take the right decision and escape.

How exactly the fishes communicate in a swarm remains a secret. Scientists have discovered the basic rules but they cannot fully explain the decentralized behavior of swarms:

- how do the fishes know where to go;
- how do the "ten" (5%) get the instructions for guiding.

In his book "Intelligence of the swarms" Peter Miller [7] describes other examples of swarm intelligence. Besides fishes he also discusses research on bees, termites, ants and starlings. They all have in common that a single living thing: a fish, a bird, an ant or a bee is quite stupid, but together they do great things. (We can build an analogy to our brain. A single brain cell is quite stupid but all together are quite smart.)

Termites are able to construct buildings that are in relation to their size enormous and related to size much bigger than men made buildings although human beings are intelligent and have advanced technologies at hand. Furthermore we have to consider that they build complete eco-systems including food production, ventilation, temperature regulation and so on. Because they care for the overall well being of the swarm.

Termites build a dynamic system and follow simple rules, like put your crumb of earth close to another crumb of earth. There is a constant exchange between the "swarm" and their environment and finally they achieve perfect living conditions. If the building gets damaged, the termites restore within a short time. They have flexible labor division and the single termite has to work in the position that is used at time. Catastrophes can damage the system but not destroy it. The swarm adapts to the change and reorganizes the system just by following their simple rules.

Scientists observed the decision making process of bees, which are looking for a new place to live after the swarm has split. The collective decision process consists of several steps: part of the bee swarm flies out and looks for options. When a bee found a place, she goes back to the others and dances. The duration and intensity of the dance represents her opinion on the quality of the place. When all bees are back the swarm decides on the quality of the dances where to live. One important characteristic is, that the bees act independently and are not influenced by the others.

Another phenomenon I want to mention are the enormous masses of starlings in Rome. They come every year and perform spectacular dances in the sky. Andrea Cavanga has observed the starlings for many years. With different cameras he filmed their performance again and again until he had enough material for his analysis. He used computer simulation to find out how they relate and communicate to each other. He came to the conclusion that each starling has fifteen to sixteen birds in its field of vision. But a single starling pays attention only to six or seven birds, which are close to him on the left and right side [8].

It is still an unknown secret, why the starlings do their artistic flights every night form November to March: If they do it just for fun or if they want to escape the falcons. Nobody knows.

Summing up, the basic characteristics of swarms are:

- Self-organization
- Simple rules
- Collective vigilance
- Resistance to failure
- Independence of meanings
- Diversity of skills
- Orientation/Adaption

2 Difference to Human Beings

Hans Hass [9] was one of the first behavioral scientist investigating the behavior of fish and he was one of the first under water filmmakers. He transferred his insights from the underwater world to the market. He wrote a book about his findings with the title "The Shark in Management". And later on he worked as consultant for companies.

He stated that human beings differ from fish and other animals by their intelligence. Other than animals they are able to build great tools, which help them to multiply their production capacity, like nowadays computers, robots and so on. To market the multiplied production money was introduced. In his opinion the introduction of money was the start point for many problems. (Fishes have no money!) Money became the "must have". The more money an individual owes the greater the power and the status. Like sharks many business people are hunting for money and take advantage when possible. Fraud and corruption are now part of our reality.

Human beings care for their personal advantage and not for the well being of the society. The environmental problems can be seen as a result of the activities of the sharks in management.

The experiments of Jens Krause have proved that human beings can organize themselves in a swarm, when they follow the rules. Unlike fishes human beings don't have the instinct to serve the community. They think about their personal benefit.

We can observe that with public transport. Everybody wants to get on the train and conquer a seat and therefore tries to get on the train as soon as possible.

Only for Mothers with little kids and old people the crowd respects the rejection zone.

Companies use a hierarchical structure to organize the crowd. The communication is basically top down. The president decides and the decisions are passed through the different levels to the employees. Actually a lot of knowledge, information and expertise get lost in the communication processes from one level to another or from one silo to another. Anyhow this organization structure seemed to work well in the past (Fig. 9).

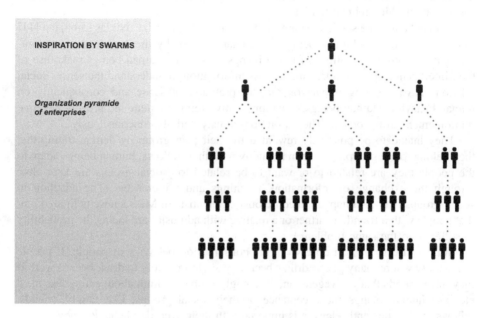

INSPIRATION BY SWARMS

Organization pyramide of enterprises

Fig. 9. Hierarchial organization structure of enterprises

More and more companies realize that they need to make better use of the skills, expertise and knowledge of their employees to be able to compete in the future. (Many years ago the president of a medium sized company commented that his company would be a smart company, if his company would knew, what his company knows. He was one of the first to build an intranet to make the knowledge of his employees accessible to all employees.)

The hierarchical structures of the past don't fit for the future. They do not work at all for companies that want to move to level five.

The communication from the bottom to the top is complicated in hierarchical structures. Mr. Gore realized that when he was an employee at Dupont [10]. He invented the fibre Gore-Tex and wanted to sell his invention to Dupont. He got lost in the hierarchy, nobody listened to him properly and nobody was interested in his project for whatever reason. Therefore he decided to build his own company – but without hierarchies. He created a company out of different self-organizing teams no bigger than 150 employees. If the company grows a new unit is installed. The employees have two rules: make money and have fun. The reputation of the employees is based on their performance and not on status.

Under certain conditions (team work in production, groups with special interests, spontaneous initiatives for help in case of catastrophes etc.) human beings have confirmed the advantages of self-organizing behavior. Contrary to swarms human beings cannot communicate as efficiently and transmit information to the whole group. Self-organization of people has a limit to size. When human beings organize themselves they need to interact with each other verbally. In groups no bigger than 150 human beings everybody gets to know each other and can establish direct communication without hierarchical borders.

Human beings are social creatures and they want to socialize. Niklas Luhmann [11] said social systems – human beings – are characterized by the exchange of communication and senses! Contrary to living beings in swarms human beings make use of their intelligence. They request and discuss information to understand the sense. Social relations have a strong impact on the interpretation of sense and consequently on human behavior. Human beings strive for positive reinforcement and prevent negative reinforcement from society. They do it consciously and also unconsciously.

They intend to be positively rewarded by their peer group by demonstrating that they belong to the group. To live in harmony with their fellows, human beings adapt to the people they are related to or want to be related to. Social groups are kept alive through the exchange of information, meanings and preferences. (The adaption to social groups may also happen unconsciously. Scientists in Massachusetts figured out, that people with a friend, a partner or a relative with adiposity are facing the probability to suffer from the same problem.)

Similar effects we can observe in the consumption behavior of people. If people join a party where many guests drinks beer they are more likely to drink beer as well. If they have friends that are vegetarian, they might probably think about eating less meat etc. For human beings the acceptance of their social groups, like family, friends, fellows, colleagues and relatives is important. In their everyday behavior they tend to swim with the stream, because they think that is a safe way.

In the business environment swimming with the stream can be a disaster. For example: the hedge fund analysts in 2008 were following the main stream, because they found it less risky to do what the others do. They were afraid to be blamed in case they act differently and fail. In the end they all failed. The financial crises showed that swimming with the stream does not prevent failure and when they group fails, the group fails great.

3 Learning from Swarm Behavior

- Self-organization:

  Build groups without hierarchies and invite agile, enthusiastic and motivated members.
  Limit the amount of participants.
  Specify the objective.
  Care for the well being of all stakeholders – the whole swarm (inside and outside).

- Diversity:

  Combine individuals with diverse expertise, skills, knowledge and experiences (because diversity is highly important to generate good ideas –
  the more diversity of knowledge, the better the quality of the solution.)

- Interaction:

  Set a few rules to structure the interaction of the group members.

- Independence:

  Group members act independently from each other.
  Diverse opinions and ideas are wanted.

- Be yourself:

  The group members truly represent themselves and stick to their opinion and do not imitate others.

- Collective vigilance:

  Group member are open and alert to new information.

- Interaction with the environment:

  The group interacts and co-creates an exchange with stakeholders outside the company and with the environment (inside out – outside in).

- Support idea generation:

  Provide a creative environment to stimulate the crowd.

- Dynamic system:

  React to changes in the environment.

- Selection process:

  Develop criteria and an effective process to reduce the generated ideas and options

- Decentralized control:

  Accept the decisions taken by the team.

- Rejection zone (Attraction, orientation, rejection):

  Consider that the variation of the rejection zone changes the behavior of the group, like time pressure, change of environmental conditions, different context etc.

- Management of uncertainty:

  Acting like a swarm helps to manage uncertainty, a swarm is resistant towards failure, wrong decisions of a few are balanced by the group.

4 Examples

New knowledge is generated faster and faster and the Internet gives everybody access to new knowledge and also to new services. Consequently the lifestyle of people is changing. Access to use becomes more important than possession. The Internet makes new services possible, like sharing, bartering and co-creation of products and services. The use of new technologies provides so many completely new options, that companies need appropriate methods, structures and procedures to deal with it.

Swarm intelligence is already part of our everyday life. Internet, smart phones, social media, apps have changed our behavior: One example is car sharing, like for example car2go [12]. The management of the fleet is self-organizing by its software. Users have access to mobility through an app and a membership card. In many cities these cars are always around and accessible.

Another example is the use of the so-called "big data".

Telephone companies receive signals from activated mobile phones and use the "big data" to communicate information on traffic jams in time to car drivers. With "big data" it is now possible to foresee twenty to thirty minutes in advance when people accumulate to a dense crowd and will probably be in danger shortly. Hopefully accidents like on the hadj and love parade can prevented in the future.

Companies that are thinking ahead act different to solve future problems and develop appropriate strategies: Working in silos of the hierarchical structures is not the way to deal with the future challenges. The growing complexity, the more specific expectations of the customers and the growing influence of the customers demand new approaches. The time of vertical operation is replaced by horizontal co-operation (Fig. 10).

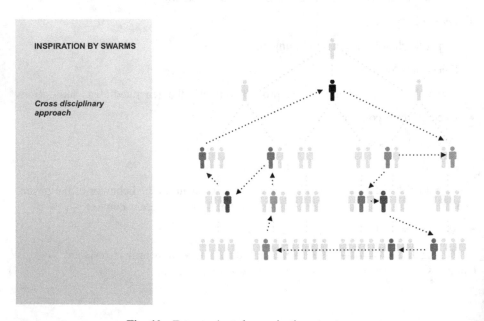

Fig. 10. Future oriented organization structures

The methods and strategies of design thinking are very popular, because their intention is to work across silos, across disciplines and across the company. Design thinking involves all stakeholders like suppliers, traders and most of all Users.

Now the silo workers got stuck with their business as usual and designers have a great time. The design awareness in companies has increased a lot. Design is been taken serious and is discussed on board level. The philosophy has changed from silo thinking to cross-disciplinary approaches with the task to create ideas, build a proto-type, test it and fail fast. Fast failure doesn't affect the company, like the wrong decision of a single fish doesn't affect the swarm. Late failure can damage a company, like a bad decision of a swarm can destroy the swarm.

Failure is no longer seen as a taboo or something very negative instead it is seen as a rich source for learning.

Companies are still very much concentrated on their mission, their values, their products, their distribution channels, their marketing and communication but not that much in the interaction with the ones the are depending on: The end-users!

A remarkable fact on the International European Design Management Congress 201 in Amsterdam was the observation that there is a change in the mindset of com-panies. Philips is a leading company in design research and in design strategies. The new aim of the company is: Improve the life of three billion people! People first – that is a new way of thinking to shape the philosophy and culture of the company [13].

Another example was the brewery Heineken. They live on selling beer – the more they sell the better. From user research they have learned that moderate drinking is preferred. Therefore they started a campaign "This one is on us"! [14] In bars in Amsterdam you get a free glass of water with each bottle of beer and a special glass fixed to the bottle. The company communicates moderate drinking to support the well being of the customers. The care for the user comes into focus and corresponds to the care for the swarm. If the users are fine the Philips and Heineken will be fine as well. User research, customer insights, co-creation, living labs and open innovation are used to better exchange with the users.

Prof. Salomão from Florianopolis visited recently presented an interesting project at Wuppertal University, which made use of swarm intelligence: The Brand "Floripa".

The task was to create a regional brand for the town Florianopolis in the south of Brazil. The project is a co-creation process and integrated inhabitants, visitors, officials and businesses into the process (Fig. 11).

The creation of a regional brand is a prestige job for design agencies. Prof. Salomão and his institute decided to work differently from the usual design agency processes. They integrated the crowd into a co-creation process:

  40 decision and opinion makers were interviewed;

  21 creative events with the community;

  900 testimonials about the DNA and the purpose;

  30.000+ engaged in person and virtually;

  29 design students and professional engaged in the visual identity.

Three alternative solutions were elaborated by the design community of Florianopolis, students and professionals were working together.

Fig. 11. Co-creation of a brand carried out by Prof. Dr. Salomão Ribas Gomez at Universidade Federal de Santa Catarina in the city of Florianopolis, Brazil

The inhabitants of Florianopolis were asked to vote on the alternatives. More than 10.000 people took the opportunity to decide about their preferred solution.

I know the city quiet well and when I look at the result I can confirm that the crowd took the best decision. The selected alternative expresses the awareness of life in the city in a perfect way. The visual identity will be applied to signage systems, brochures, merchandise articles, public places and events and so on.

I am not sure, if the lord-mayor by himself would have taken the same best decision (Fig. 12).

There is a demand for new processes of co-operation and co-creation in complex processes. Another example I want to mention was elaborated by one of my doctoral students. His job is to manage the exhibitions of Mercedes Benz on international automobile fairs. Those of you who have visited such a fair may imagine that it is a very complex task.

All stakeholders involved have their special interests: the top management, the engineers, the marketing people, the designers, the people in charge of public relation, the sales department and finally the interests of visitors and prospect buyers have to be considered. The challenge is, how to manage the crowd of experts – the swarm. – They are supposed to stick to the company's values and not to their individual vanity. My doctoral student did an intense empirical research and included all stake holders and their special interests. The final outcome is a tool similar to a compass that enables him to ask the right question at the right time to the right person to strengthen the co-operation of the team. (complexity broken down to simple rules to manage the swarm).

Fig. 12. Co-creation of a brand carried out by Prof. Dr. Salomão Ribas Gomez at Universidade Federal de Santa Catarina in the city of Florianopolis, Brazil

5 Swarm Intelligence as Inspiration for a Design Strategy Project

Like a swarm a company is not a closed system. Each swarm and each company is also part of a bigger system. The collective vigilance plays an important role. The interaction with the environment is not only important for swarms it is equally important for companies and decides about life and death.

Companies realize that design thinking enables them to prepare for the future and to reach level five of the design ladder. The permanent interaction with the users and the environment seems the best way to create a sustainable future.

Wera is a medium sized company producing screwdrivers of all types in high quality, excellent functionality and extremely user friendly. We started to work with them in the summer term 2013. That time they were on level three of the design ladder and intended to climb to level four.

Together with my assistants and with our master students I elaborated a design strategy and a brand strategy for the company. The management was highly motivated. Our co-operation was good and trustful, although or because we analyzed and benchmarked the company critically and we always told them truth.

They were happy with the result and used the handbooks to implement the strategies. In the last three years the company has changed not only its appearance but also the way they work together. They work together like a swarm, they share the same DNA and follow the rules and they are enthusiastic. The investments in design have paid off. Meanwhile they won a lot of design prizes and have doubled their turnover (Fig. 13 and 14).

Fig. 13. Fair stand of Wera at "Internationale Eisenwarenmesse" in Cologne, March 2016

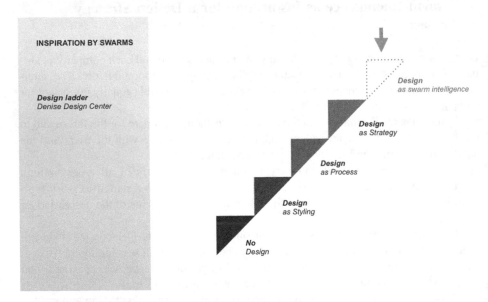

Fig. 14. Application of swarm theory to climb up the design ladder to level five

I visited their fair stand in April and I was surprised: the fair stand was totally crowded with people and the Wera employees on the fair stand were extremely busy to handle all requests of the visitors. In compare the fair stands of the competitors were quite empty. Wera runs ahead of the competitors, but the company knows very well that the most important competitors will catch up soon. Therefore they decided to improve further. Their desired goal is to become a public brand, like aspirin. (In almost ever country of the world aspirin is a synonym for a painkiller, even if it is not the original.)

They definitively want to reach level five of the design ladder!

To reach level five they need to operate differently and they need to interact with the environment directly - like swarms do. New technologies provide great options to interact direct and fast. Close and good relation to the environment and the customers will be their key for future survival. In my opinion future success will be determined by:

1. transparent interactions with users and all the other stakeholders;
2. demonstrations of environmentally and socially responsive and sustainable activities of companies.

My proposal for the process considers what I have learned of swarm behavior:

1. Change your mindset (DNA) from offering to caring

   Workshop with the top management to discuss the new aim.
   We want to figure out:
   What works well and what does not work so well?
   What are the problems?
   What are the challenges the company is facing today?
   What are the aims and objectives of the company for the future?
   Result: Clear definition of the future perspective.

2. Design thinking process

   Organization of cross-disciplinary workshops as self-organizing groups.
   We have to:
   Provide the rules how to interact and a friendly setting.
   Invite the participants, like employees of all departments, representatives of the different distribution channels and end users from different production companies.
   Guarantee diversity, independence and be yourself.
   Results: identification of existing problems, generation of ideas how to solve the problems the different stakeholder are facing right now, creation of as many ideas as possible.

3. Orientation

   We have to:
   Structure the ideas and prepare the selection process.
   Consider the aim of the company as the basis for the selection criteria.

4. Evaluation

   We have to organize a cross-disciplinary workshop to evaluate and select the most promising ideas (wisdom of the crowd, swarm behavior).

5. Testing

Rapid prototyping of selected ideas and test them in real life experiments (interaction with the environment).

6. Learn from failure and improve

Cross-disciplinary workshop to rethink the experiment and to find solution for improvement (swarm decision to balance the wrong decision of an individual).

7. Design process
8. Implementation

When the project succeeds the company will not only act as a swarm internally, the company will act as swarm in the ocean. The company will sustain shark attacks because the employees (swarm) are alert to changes in the environment and react spontaneously, they observe the behavior of other swarms (competitors) and most important they interact with users and retailers in co-creation processes.

The main motive to join a swarm is the fear to be eaten by an enemy.

The same fear exists in companies, they are afraid to be swallowed by the competition. I am convinced: When they learn from swarm behavior they will be stronger, climb to level five and have better chances for long lasting success in the market.

References

1. Shubin, N.: Der Fisch in uns: Eine Reise durch die 3,5 Milliarden Jahre alte Geschichte unseres Körpers (2009)
2. Reynolds, Craig: http://www.red3d.com/cwr/boids/
3. Krause, J., Krause, S.: Kollektives verhalten und schwarmintelligenz. In: Otto, K.-S., Speck, T. (eds.) Darwin meets Business, pp. 127–134. Springer, Heidelberg (2011)
4. Couzin, I., Miller, P.: The Smart Swarm. Penguin Books, New York (2010)
5. Krause, J.: Swarm Experiment in the Cologne Fair and reported in the TV program Quarks&Co by Ranga Yogeshwar (2007)
6. Surowiecki, J.: The Wisdom of the Crowd: Why the Many are Smarter Than the Few and How Collective Wisdom Shapes Business. Society and Nations (Abacus), Economics (2005)
7. Miller, P.: The Smart Swarm. Penguin Books, New York (2010)
8. Cavanga, A., Miller, P.: The Smart Swarm. Penguin Books, New York (2010)
9. Hass, H.: Der Hai im Management, Instinktverhalten erkennen und kontrollieren (1988)
10. DuPont – Wikipedia. https://de.wikipedia.org/wiki/DuPont/
11. Luhmann, N., Systeme, S.: Grundriß einer allgemeinen Theorie, Frankfurt am Main 1984, neue Auflage (2001)
12. The car2go Web site. https://www.car2go.com/DE/de/
13. Phillips: Company presentation at the European International Design Management Congress, Amsterdam, Netherlands (2016)
14. Heineken: Company presentation at the European International Design Management Congress, Amsterdam, Netherlands, during the reception in the Heineken Building (2016)

Regular Papers

Particle Swarm Optimization for Operating Theater Scheduling Considering Medical Devices Sterilization

Benoit Beroule[1]([⊠]), Olivier Grunder[1], Oussama Barakat[2], Olivier Aujoulat[3], and Helene Lustig[3]

[1] Univ. Bourgogne Franche Comté, UTBM, IRTES-SET, 90010 Belfort, France
{benoit.beroule,olivier.grunder}@utbm.fr
[2] Nanomedecine Lab, University of Franche Comté, 25000 Besançon, France
oussama.barakat@univ-fcomte.fr
[3] GHRMSA, Mulhouse Hospital Center, 68000 Mulhouse, France
{aujoulato,lustigh}@ch-mulhouse.fr
http://www.utbm.fr
http://www.univ-fcomte.fr
http://www.ch-mulhouse.fr

Abstract. The operating theater scheduling problem is one of the main hospital sector issues of today's world. Indeed, numerous papers dealing with this subject may be found in the literature. However, the synchronization between the pharmacy (providing the surgical devices and medicines) and the operating theater is rarely studied. Nevertheless, the importance of the pharmacy keeps growing because of the creation of numerous hospital groups composed of several hospital complexes sharing a central pharmacy. In this paper, we focus on the sterilization cycle of the surgical devices to provide operating theater scheduling methods taking into account pharmacy issues. We present exact methods with a mixed integer linear programming model to determine optimal schedules as well as approximate solutions with a particle swarm optimization based method to solve the most complex cases. These modelings provide interesting schedules using few quantities of surgical devices boxes even when considering many procedures. With this study we hope to lay the foundations of a transverse logistics unifying the operating theater and the pharmacy in a multi-site context.

Keywords: Optimization · Health care · Particle swarm optimization · Operating theater scheduling

1 Introduction

The health-care sector must adapt to any economic situation more than any other public sector to ensure the continuance of every care services. To continue operating at optimum efficiency, lots of hospital complexes merge to create

© Springer International Publishing AG 2016
P. Siarry et al. (Eds.): ICSIBO 2016, LNCS 10103, pp. 41–56, 2016.
DOI: 10.1007/978-3-319-50307-3_3

groups sharing a common logistics in terms of transportation, medical products, medical devices and more.

In this work, we study the case of a real pharmacy in a new hospital group which is intended to become the central pharmacy. In practical terms, it implies the centralization of the medical devices and drugs storage as well as the sterilization of the medical devices. This study is particularly focused on this second aspect. The paper proposes to initiate the creation of a transverse logistics between the pharmacy and the operating theater to develop scheduling methods that take into account the local logistics of these two entities [1]. Such an approach implies defining a new performance criterion. More importantly, it should be possible for this work to be extended to the multi-site context [18]. To the best of our knowledge, this particular aspect of the operating theater scheduling has never been studied before. A multi-site context implies an important amount of surgical procedures to schedule each day. In order to solve the considered operating scheduling problem with numerous data, we developed a particle swarm optimization-based method. This paper details the complete PSO method used as well as an empirical selection and a discussion on the parameters.

There are numerous papers which study the operating theater management. [10] proposed to use Bin-Packing methods derived from the industrial sector, and fuzzy constraints to maximize rooms occupancy rate in a multidisciplinary operating theater scheduling problem.

[14] developed a tabu search adapted to the hospital context to manage the operating rooms considering the recovery rooms availability. Other authors like [13] have chosen to propose scheduling tools based on linear models to minimize the operating cost of the hospitals. [27] studied the importance of reserving a dedicated operating room in case of emergency to improve responsiveness when treating non-elective (urgent) patients. The performance criterion selected to evaluate room scheduling procedures is an important aspect of operating theater scheduling. It may affect the approach used to solve the problem. [4] identify the eight main performance criteria detailed in the literature which are: waiting time, throughput, utilization leveling, makespan, patient deferrals, financial measures and preferences. For instance, lots of studies are based on avoiding under-utilization and over-utilization of the operating rooms by elaborating operational research methods [8–10]. On the other hand, [7] implemented a linear programming model to maximize the costs induced by the operating theater to determine the worst possible case. [2,3] consider the NP-hard combinatorial optimization problem of surgical case sequencing with a multi-objective function and proposed several methods to solve it in a freestanding ambulatory unit context by determining the order of the patients. Indeed most of the papers re centralized on the patients rather than the materials, for instance, [19] proposed scheduling solutions to decrease the patient makespan that is to say the time elapsed between the first patient arrival and the last patient departure within the considering time window.

This paper will be divided into several parts. In the second part, the context of the study will be presented and the sterilization service relative problems will be detailed. In the third part, the modeling method will be described as well as a mathematical model. The Fourth part presents the particle swarm optimization based algorithm and a parameter determination method. The results are discussed in the fifth part to compare the different methods in terms of solution and computation time. At least, we will conclude on the best method to determine optimal scheduling in this context and suggest some projects.

2   Studied System

The considered study takes place in a hospital pharmacy which needs to face up to the new economic situation. In this section, the current functioning of the pharmacy and its future development is presented and the main encountered logistic problems are highlighted.

2.1   The Multi-site Context

The studied hospital is a part of a recently formed hospital group. Consequently, the logistics of its pharmacy must be reorganized to become the central pharmacy of the group. Concerning the sterilization service, this major modification means that every hospital of the group will send their own medical devices to this pharmacy to be sterilized then sent back to the hospital (or another hospital of the group). In this context, we propose methods to avoid congestion in the sterilization service. We assume that working on a new global surgical procedure scheduling process which takes into account the resources of the pharmacy may be relevant. Our final purpose is to provide methods which can be implemented into decision making tools to determine surgical procedure schedules concerning an entire hospital group.

2.2   Sterilization Service

The sterilization service of the pharmacy is responsible for the washing, the sterilization and the repackaging of the medical devices into boxes (we simply denote them "boxes" in the remainder of the paper). When a surgical procedure occurs, the corresponding box follows a precise cycle (summarized in Fig. 1). Before the surgical procedure starts, the appropriate box (or boxes) is (are) sent from the sterilization service to the operating theater with a vehicle (automated or not) or sometimes directly via a human agent. The box is brought to the appropriate operating room to be used during the procedure (it can be stored in the service before the operation starts if necessary). When the surgical procedure is over, the medical devices are pre-disinfected by immersing them into a disinfectant liquid during twenty minutes (average duration). Then the instruments are repacked in the right box and stored in a common dedicated zone before being sent back to the sterilization service of the pharmacy. When the boxes are received by the service, they follow several steps to be sterilized. First the medical devices are divided among automatic washers to be cleaned. When the washing program is

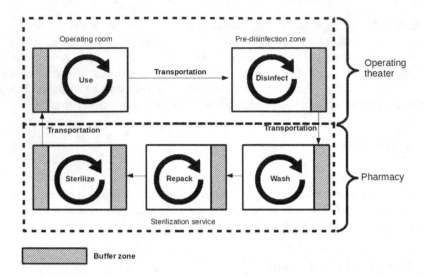

Fig. 1. Medical devices cycle

over, the materials must be repacked in the corresponding boxes. This operation may be long or less and complex depending on the number of instruments (which might exceed one hundred) and the complexity of the placement in the box. The repacked boxes are then stored in a dedicated buffering zone. When the adequate number of boxes are stored, they are put into autoclaves (sterilization machines) to be sterilized. This process lasts more than an hour. When it is over, the boxes are stored while their temperature drops. After this step, the sterilization process is over and the boxes are available. They may be sent to the operation theater to continue the cycle when needed. It is important to notice that this cycle may be interrupted during precise moments (before/after washing or before/after sterilizating) but under no circumstances during the surgical procedure.

2.3   Problematics

Currently, there is no 'transverse logistic' between the operating theater and the pharmacy in the studied hospital complex. It means that the surgery operations are scheduled without taking into account the pharmacy resources. This method presents two main problems.

First, each box is dedicated to a precise surgery operation type. As a consequence, there must be enough boxes for each surgery procedure to respect any possible planning. A better scheduling method may reduce the number of needed boxes implying an important saving in a multi-site context.

Furthermore, the sterilization service must deal with important bursts in activity when too many procedures end during a short time period (or because of a transportation problem) while sometimes, no box is returned to the operation theater during a long time period. By balancing the workload, we aim to improve the efficiency of the service.

3 Modeling

We first identify the difficulties relative to the connection between the pharmacy and the operating theater. We notice that the problem may be compared to a standard scheduling problem in industrial context with some particularities. Consequently, we start our study by using industrial models. Then we implement two solving methods. On the one hand, we developed a mixed integer linear programming model to obtain optimal schedules (minimizing boxes utilization), the main weakness of the method is a prohibitive computation time when treating important size problems. To improve the methods, On the other hand we develop a particle swarm optimization based-model in order to provide schedules even when considering numerous data.

3.1 Operating Theater Scheduling

The Operating theater scheduling is a complex problem which deals with numerous variables (human resources, material resources, operation rooms, time window, skills...). In this preliminary study, we simplify the problem and then will consider afterward the constraints implied by the other operating theater aspects. This leads us to the following assumptions:

– Each surgical procedure is already assigned to a surgical team.
– The human resources are always available.
– All operating rooms are open simultaneously.
– All patients are ready for the surgical procedure.
– The non-elective patients (urgency) are not considered.
– The problems relative to the recovery room are not considered.
– No transportation problem may occur.

This scheduling problem shares similarities with classical industrial scheduling problem. Hence, it may be interesting to use industrial optimization methods with proper modifications to identify the problem. [12] proposed a nomenclature adapted to scheduling models in industrial context, this nomenclature identifies a scheduling problem by defining three parameters: α, β and γ.

– α is composed of two sub-parameters $\alpha = \alpha_1\alpha_2$.
   • α_1 represents the layout of the actual shop (flow-shop, job-shop, open-shop...).
   • α_2 indicates the number of machines in the shop and if this number is fixed or variable.
– β represents constraints which may be applied to the problem (due dates, ready dates, deadlines, preemption, precedence relation...).

– γ represents the selected performance criterion. The choice of this parameter is crucial to determine the aspect that must be improved by the scheduling.

There are a lot of papers that compare the operating theater to an hybrid flow-shop [11,14,15,23]. These studies assimilate the patients to jobs and the operating theater steps to machines. In this work, we concentrate on surgical devices. Consequently, we assimilate surgical procedures to jobs and boxes to machines. In this context, according to the nomenclature, the problem must be compared to a "RI/ $prec,pmtn$/Boxes utilization minimization" problem.

$\alpha_1 = R$ means that the operations are parallel and unrelated. Despite the fact that the machines represent the same type of box, the time needed for a job depends on other parameters (the patient, the surgeon...).

$\alpha_2 = I$ means that the number of machines (or boxes) is fixed.

$\beta = prec,pmtn$

$prec$ means 'precedence'. We separate each job into two sub-jobs. The first sub-job represents the operating theater and transportation part and the second one represents the sterilization service part. Hence we add a precedence constraint between these two sub-jobs (operating theater first then sterilization).

$pmtn$ means 'preemption'. As said before, the sterilization sub-jobs may be pre-empted.

$\gamma =$ "Boxes utilization minimization". There is no pre-existing parameter which correctly represents our scheduling problem performance criterion. Indeed, we try to minimize the number of machines (boxes) which will be used and this kind of criterion is rarely considered in the literature.

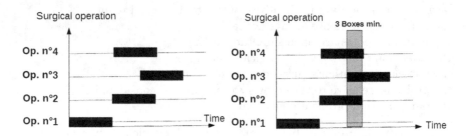

Fig. 2. Example of a four-surgical operation planing

To model the operating theater scheduling problem considering the sterilization step, each prescribed surgical procedure is represented by a job j the duration of which corresponds to the time needed to perform a complete cycle of the corresponding box (see Sect. 2.2). The job starts simultaneously with the corresponding surgical procedure. Therefore two distinct surgical procedures can share a same box only if the corresponding jobs do not overlap. It is then possible to represent a complete planning for a given procedure type (sharing the same type of medical devices). This planning may be evaluated by calculating the number of boxes needed to respect it, which is easily identifiable on a graphic representation, an example of a four-surgical operation planing is presented on Fig. 2. This example presents the time repartition of the four surgical operations, the black strip represent the time during which the corresponding box will be unavailable, it corresponds to a complete box cycle (Fig. 1). In this condition, the best planning is the one with the lowest evaluation (lower amount of boxes needed to be respected).

3.2   Mixed Integer Linear Programing Model

As seen before, the scheduling problem considered here is difficult to identify because of its quite unusual γ parameter. This is why the problem will not be directly treated as a scheduling problem. Consequently, we propose a mathematical model based on the bin packing problem. The bins are replaced by the boxes and the objects by the surgical procedures then we added new time relative constraints. We consider here that each surgical procedure implies the creation of a job representing the corresponding box utilization cycle (see Sect. 2.2)

the parameters of the model are defined:

- n: number of surgical procedures.
- P_i: processing time of the job i in the operating theater and transportation.
- S_i: processing time of the job i in the sterilization service.
- S^- opening date of the sterilization service.
- S^+: closing date of the sterilization service.
- P^-: opening date of the Operating theater.
- P^+: closing date of the Operating theater.
- T: duration of a sterilization working day $(T = P^- - P^+)$

then the relevant indexes:

- i: denote the job/procedure index $1 \leq i \leq n$.
- j: denote the box index $1 \leq j \leq n$.
- k: denote the day of the week index $1 \leq k \leq 5$.

Then the following decision variables are introduced:

Primary

- $x_{i,j,k}$: a binary variable which is equal to 1 only if the operation i uses the box j during the day k.
- y_j: a binary variable which is equal to 1 only if the box j is used for at least one job.

Secondary

- t_i^-, t_i^+: date of the beginning and the end of each job i respectively.
- a_i, b_i, c_i: binary variables used to create time relative constraints (see Eqs. (5) to (8))

We define the objective function as follows:

$$Minimize \sum_{j=1}^{n} y_j \tag{1}$$

Subject to the constraints:

$$\sum_{i=1}^{n} x_{i,j,k} \times (P_i + S_i) \leq T \times y_j \quad \forall j \in [\![1,n]\!] \quad \forall k \in [\![1,5]\!] \tag{2}$$

$$\sum_{j=1}^{n} \sum_{k=1}^{5} x_{i,j,k} = 1 \quad \forall i \in [\![1,n]\!] \tag{3}$$

$$x_{i_1,j,k} \times x_{i_2,j,k} \times (t_{i_1}^- - t_{i_2}^+) \times (t_{i_1}^+ - t_{i_2}^-) \geq 0 \quad \forall i_1, i_2, j \in [\![1,n]\!] \quad \forall k \in [\![1,5]\!] \tag{4}$$

$$T^- + a_i \times T \leq t_i^- \leq T^+ + a_i \times T \quad \forall i \in [\![1,n]\!] \tag{5}$$

$$T^- + a_i \times T \leq t_i^- + P_i \leq T^+ + a_i \times T \quad \forall i \in [\![1,n]\!] \tag{6}$$

$$S^- + b_i \times T \leq t_i^- + P_i \leq S^+ + b_i \times T \quad \forall i \in [\![1,n]\!] \tag{7}$$

$$S^- + c_i \times T \leq t_i^+ \leq S^+ + c_i \times T \quad \forall i \in [\![1,n]\!] \tag{8}$$

$$(k-1) \times T \times x_{i,j,k} \leq t_i^- \leq 5 \times T \times (1-k) \times x_{i,j,k} \quad \forall i, j \in [\![1,n]\!] \quad \forall k \in [\![1,5]\!] \tag{9}$$

The purpose here is to minimize the number of used boxes (Eq. (1)). The Eqs. (2) and (3) are classical bin packing equations, they ensure the jobs to be assigned to one precise day and one precise box. Equation (4) prevents the jobs from overlapping when using the same box. The equations from (5) to (8) define the time windows of the operating theater (when the surgical procedures may start or end) and of the sterilization service (when the sterilization process may start or end). Finally, Eq. (9) forces the dates t_i^1 to be within the right time window depending on the day of the corresponding surgical procedures. Note that the Eq. (4) can be easily linearized to obtain a MILP model.

4   Particle Swarm Optimization

The particle swarm optimization (PSO) is a parallel evolutionary computation meta heuristics invented by Kennedy and Eberhart [16,17,24] which is based on bird-flocking and fish-schooling. Particles are created in the solutions space and share information to move and converge towards best solutions. Numerous papers deal with PSO improvements or practical applications [20–22]. A specific method has been defined to choose PSO parameters in order to improve convergence rate and identify each parameter utility [5]. Indeed, the parameters greatly affect the solutions consistency. Consequently, some papers studied their impact in a mathematical [26] or empirical [25] way. In this section the particle swarm optimization base algorithm used to solved the problem is described. This approach is necessary when solving problem with large amount of surgical operations.

4.1 Particle Swarm Optimization Modeling

Implementing a PSO algorithm implies determining the modeling of the particles which will explore the solutions space. Our purpose is to determine a one week surgical procedures planning by determining the starting date of each operation. Furthermore, the duration time of a procedure is not a decision variable and may mainly depend on the patient physical characteristics, the pathology type or the surgeon habits. In these conditions, the starting dates are sufficient to establish a complete planning with approximate duration times.

We define the following PSO relative modeling parameters (some of them will be detailed afterward).

- m: amount of particles generated for the PSO algorithm.
- l: amount of steps performed by the PSO algorithm.
- X_j^k: position vector of the particle j at step k.
- V_j^k: velocity vector of the particle j at step k.
- L_j: best solution found by the particle j.
- G_j: best solution found by the particles of particle j's neighborhood.
- D: neighboring distance.
- $d_{i,j}^k$: distance between particle i and j at step k.
- ω: inertia factor.
- ϕ_1: personal memory factor.
- ϕ_2: common knowledge factor.
- Ω_1: solutions space.
- Ω_2: computation space.
- r_1^k, r_2^k: vectors of random generated float from 0 to 1.
- Σ: set of scenarios.

Hence, each particle is represented by its position and velocity. The position is a n-tuple as shown in Eq. (10).

$$X_j^k = (t_1^-, t_2^-, ..., t_n^-) \tag{10}$$

With this modeling, the particles progress in an n dimensional space. A movement along a dimension i represents a modification of the corresponding starting date t_i^-. To initialize the PSO, m particles will be generated with random starting dates distributed during the coreponding week and random initial velocities V_j^0. m must be big enough to create a set of particles covering the entire solution space. At each step k, a particle represents a particular solution according to its position in the solution space.

During each step of the PSO algorithm, the particles will communicate to share information and update their own positions according to their own knowledge and the common knowledge of the best solution. The details of the new position computation are given in Eqs. (11) and (12) [17].

$$V_j^{k+1} = \omega V_j^k + \phi_1 r_1^k (L_j - X_j^k) + \phi_2 r_2^k (G_j - X_j^k) \tag{11}$$

$$X_j^{k+1} = X_j^k + V_j^{k+1} \tag{12}$$

$$d_{i,j}^k = \sqrt{\sum_{\eta=1}^{m}(X_j^k(\eta) - X_i^k(\eta))^2} \qquad (13)$$

L_j and G_j represent the position vectors of the best solutions found by the particle j and by the particles in its neighborhood respectively. They are updated at each step if needed. The neighborhood of a particle j is a set composed of every other particles i which distance $d_{i,j}^k$ is lesser than D, the distance between two particles is given by Eq. (13). Here the Euclidian distance is used in order to create neighborhood composed of slightly similar solutions, that is to say solutions with some small time offsets. Moreover, it is important to note that here the distance between two solutions must be expressed in time units (commonly hours). ω represents the system global inertia. A high inertia value implies a better solution space exploration at the expense of the convergence speed. ϕ_1 and ϕ_2 represent the personal memory factor and the common knowledge factor respectively. If ϕ_1 is set to a high value, each particle will be more attracted by its own best already visited position L_j. If ϕ_2 is set to a high value, each particle will be more attracted by the best already visited position among every visited positions of every particles G_j.

When the final step is reached, the solution corresponding to the best visited position among every particles is considered as the PSO algorithm output. l must be large enough to allow the particles to converge toward one or several extrema, but not too big to prevent the machine from prohibitive computation time.

Among other factors, the PSO efficiency depends on the solution space topology and the fitness function behavior. Here we define the solution space Ω_1 as all possible dates combinations in a week (Eq. (14)).

$$\Omega_1 = \{(t_1^-, t_2^-, ..., t_n^-) | \forall i \in [\![1, n]\!], 0 \le t_i^- \le 5 \times T,\ T_- \le t_i^- \bmod T < T_+\} \quad (14)$$

In this scheduling problem, the fitness function evaluates the number of needed boxes to respect a given schedule. The problem is that Ω_1 is a discrete subset of $\mathbb{R}^n$, this topology particularity prevents the particles from moving in a continuous way. To improve the PSO efficiency, we consider a new space, Ω_2 (continuous subset of $\mathbb{R}^n$), that will be called the "computational space" ((15)).

$$\Omega_2 = \{(t_1^-, t_2^-, ..., t_n^-) | \forall i \in [\![1, n]\!], 0 \le t_i^- < 5 \times (T_+ - T_-)\} \qquad (15)$$

There is a bijective function from Ω_1 to Ω_2 (Eqs. (16) and (17)) to translate the straight forward readable solution from Ω_1 to Ω_2 where the computation is easier. When the computation is over, the solutions may be translated back from Ω_2 to Ω_1 (Eqs. (18) and (19)).

$$f: \qquad \Omega_1 \to \Omega_2$$
$$(t_1^-, t_2^-, ..., t_n^-) \mapsto f((t_1^-, t_2^-, ..., t_n^-)) = (t_1'^-, t_2'^-, ..., t_n'^-) \qquad (16)$$

$$t_i'^- = (t_i^- - T^-) - \left\lfloor \frac{d_i}{T} \right\rfloor \times (T + T^- - T^+) \qquad (17)$$

$$f^{-1}: \qquad \Omega_2 \to \Omega_1$$
$$(t_1'^-, t_2'^-, ..., t_n'^-) \mapsto f^{-1}((t_1'^-, t_2'^-, ..., t_n'^-)) = (t_1^-, t_2^-, ..., t_n^-) \qquad (18)$$

$$t_i^- = (t_i'^- + T^-) + (T + T^- - T^+) \times (t_i'^- \bmod (T^+ - T^-)) \qquad (19)$$

4.2 Determining Best Parameters

In order to improve the PSO efficiency, we perform a sensibility analysis on the ω, ϕ_1 and ϕ_2 factors on the solution provided by the PSO algorithm. We therefore implement an algorithm to find the best parameters values (Fig. 3). The purpose of this algorithm is to compute the best solution fitness average for any parameters triplet (ω, ϕ_1, ϕ_2) considering a scenario σ provided as inputs. To represent the algorithm outputs, a 3-dimensional data structure $F_\sigma(\omega, \phi_1, \phi_2)$ is set and initialized to 0. The PSO algorithm is then performed several times (depending on $NbIter$) with the given parameters and $F_\sigma(\omega, \phi_1, \phi_2)$ is updated at each iteration by adding the fitness of the best found solution (provided by $PSOBestSolutionFitness(i, j, k, \sigma)$). When the final iteration is reached, $F_\sigma(\omega, \phi_1, \phi_2)$ is divided by $NbIter$ to obtain an average result then another parameters triplet can be evaluated. In this study, this evaluation is maid for any triplet $(\omega, \phi_1, \phi_2) \in P$ (define in Eqs. (20) to (23))

```
const
     NbIter: Integer;
     Sigma: Scenario;
     omegaStart, omegaStep, omegaEnd: Real
     phi1Start, phi1Step, phi1End: Real
     phi2Start, phi2Step, phi2End: Real
var
     i := omegaStart; j := phi1Start; k := phi2Start;
     it: integer;
     Fs: Real 3 dimensional data structure;
begin
     repeat
          repeat
               repeat
                    it := 1;
                    Fs(i,j,k) := 0;
                    repeat
                         Fs(i,j,k) := Fs(i,j,k) + PSOBestSolutionFitness(i,j,k,Sigma);
                         it := it + 1;
                    until it > NbIter
                    Fs(i,j,k) := F(i,j,k) / NbIter;
                    k := k + phi2Step
               until k > phi2End
               j := j + phi1Step
          until j > phi1End
          i := i + omegaStep;
     until i > omegaEnd
end
```

Fig. 3. PSO best parameters evaluation algorithm

$$P = P_\omega \times P_{\phi_1} \times P_{\phi_2} \tag{20}$$

$$P_\omega = \{\omega \in \mathbb{R} | \exists i \in \mathbb{N}, \omega = \omega_{start} + i \times \omega_{step}, \quad \omega \leq \omega_{end}\} \tag{21}$$

$$P_{\phi_1} = \{\phi_1 \in \mathbb{R} | \exists i \in \mathbb{N}, \phi_1 = \phi_{1\,start} + i \times \phi_{1\,step}, \quad \phi_1 \leq \phi_{1\,end}\} \tag{22}$$

$$P_{\phi_2} = \{\phi_2 \in \mathbb{R} | \exists i \in \mathbb{N}, \phi_2 = \phi_{2\,start} + i \times \phi_{2\,step}, \quad \phi_2 \leq \phi_{2\,end}\} \tag{23}$$

Therefore, we may define a set of scenarios $\Sigma = \{\sigma_1, \sigma_2, ..., \sigma_\nu\}$. The best parameters obtained by the algorithm (3) for a set of scenarios Σ are given in Eq. (24)

$$(\omega_{best}, \phi_{1\,best}, \phi_{2\,best}) = arg\,min(\sum_{\sigma \in \Sigma} F_\sigma(i, j, k)) \tag{24}$$

Here we define the ranges of value for each parameter with:
$\omega_{sart} = \phi_{1\,start} = \phi_{2\,start} = 0.2$,
$\omega_{step} = \phi_{1\,step} = \phi_{2\,step} = 0.2$,
$\omega_{end} = \phi_{1\,end} = \phi_{2\,end} = 2.0$,
to obtain the empirical best parameters of the Eq. (25).

$$(\omega_{best}, \phi_{1\,best}, \phi_{2\,best}) = (0.2, 1.6, 1.6) \tag{25}$$

We do not assure that the previously determined parameters are the best choice to converge toward the best solution but we assume they are an interesting alternative considering the fact that only 2 h (with *NbIter* = 50) were needed to compute them. Let us consider the consistency of this result. A theoretical approach leads to define the PSO factors by the equations $\phi_1 = \phi_2 = \phi$ and $\phi = \omega \times (2/0.97725)$ [6], this is why we first decided to use the parameters $(\omega, \phi_1, \phi_2) = (0.8, 1.64, 1.64)$. From the empirical results of testing, two observations can be made. First $\phi_{1\,best} = \phi_{2\,best} = 1.6$. However the inertia factor $\omega_{best} = 0.2$ is smaller than the expected value (0.8). To understand this result, let us remember the impact of this parameter on the global system. The inertia factor represents the particles capacity of "quickly" changing their directions, therefore, the bigger the inertia factor is, the more the solution space is explored (but the convergence rate may decrease). Nevertheless, the solution space of the current problem contains several non-neighboring optimal solutions (for instance, inverting two surgical procedures of same duration provides an other solution with identical fitness). Consequently, the exploration of the entire solution space is not crucial, hence the inertia factor does not need to be set to a high value in this context.

5   Experimentation

A set of representatives scenarios Σ composed of different number of procedures is defined. These scenarios are inspired by real data from hospital context, some of them represent real cases and are extended to create the other ones. Proceeding in this manner allows to obtain a set of scenarios Σ with a large panel of surgical procedures. Here Σ contains 32 scenarios composed of 6 to 37 surgical

Table 1. Number of boxes needed to respect each scenario depending parameters value and MILP model

Scenario	Procedures	PSO_1			PSO_2			MILP	
		Fit	Time(s)	Error	Fit	Time(s)	Error	Fit	Time(s)
1	6	2.00	7.9	0.0%	2.00	7.9	0.0%	2	0.09
2	7	2.00	9.0	0.0%	2.00	8.9	0.0%	2	0.3
3	8	2.00	10.0	0.0%	2.00	10.0	0.0%	2	0.7
4	9	2.01	11.1	0.5%	2.01	11.1	0.5%	2	2.17
5	10	2.83	12.3	41.5%	2.64	12.4	32.0%	2	1.23
6	11	3.00	13.8	0.0%	3.00	13.8	0.0%	3	26.64
7	12	3.00	15.3	0.0%	3.00	15.3	0.0%	3	34.43
8	13	3.00	16.9	0.0%	3.00	17.0	0.0%	3	10.05
9	14	3.36	18.4	12.0%	3.15	18.7	5.0%	3	84.36
10	15	3.93	20.4	31.0%	3.91	20.7	30.3%	3	80.1
11	16	4.00	22.7	0.0%	4.00	23.0	0.0%	4	113.1
12	17	4.00	24.6	0.0%	4.00	25.1	0.0%	4	36.47
13	18	4.06	26.2	1.5%	4.02	26.3	0.5%	4	152.39
14	19	4.57	28.4	14.25%	4.30	28.9	7.5%	4	210.98
15	20	4.99	30.2	24.75%	4.99	30.7	25.0%	4	58.66
16	21	5.00	32.3	0.0%	5.00	33.0	0.0%	5	143.28
17	22	5.12	34.3	2.4%	5.00	35.0	0.0%	5	78.35
18	23	5.63	36.3	12.6%	5.36	37.5	7.2%	5	495.25
19	24	5.93	38.2	18.6%	5.91	39.3	18.2%	5	558.51
20	25	6.00	40.9	20.0%	5.98	41.8	19.6%	5	591.50
21	26	6.03	43.0	0.5%	6.00	44.0	0.0%	6	808.80
22	27	6.19	45.4	3.16%	6.08	46.7	1.3%	6	820.41
23	28	6.75	47.6	12.5%	6.57	49.0	9.5%	6	2413.1
24	29	6.97	50.2	16.17%	6.99	51.9	16.5%	6	2842.1
25	30	7.07	52.5	16.8%	7.00	54.3	16.6%	6	3109.5
26	31	7.33	55.1	-	7.11	57.0	-	-	≥ 3600
27	32	7.80	58.0	-	7.56	59.8	-	-	≥ 3600
28	33	7.93	60.6	-	7.92	62.4	-	-	≥ 3600
29	34	8.03	63.3	-	7.98	65.7	-	-	≥ 3600
30	35	8.15	66.1	-	8.00	68.6	-	-	≥ 3600
31	36	8.49	69.1	-	8.22	71.4	-	-	≥ 3600
32	37	8.76	72.0	-	8.65	74.7	-	-	≥ 3600

procedures, the amount of procedures in a scenario determines its complexity and therefore the computation time needed to solve it. We evaluate the schedules provided by two different PSO algorithms.

- PSO_1 uses the classical parameters $(\omega, \phi_1, \phi_2) = (0.8, 1.64, 1.64)$. These values are standard and respect the rules prescribe in the literature [5, 6].
- PSO_2 uses the parameters determined by the best parameters determination method presenting in Sect. 4.2 that is to say $(\omega, \phi_1, \phi_2) = (0.2, 1.6, 1.6)$.

For each scenario of Σ, the operating theater scheduling problem described in this paper has been solved using the two methods PSO_1 and PSO_2 100 times to evaluate an average fitness and computation time. Then the GUROBI^{TM}solver has been used to determine the optimal solution fitness according to the Mixed Integer Linear Problem modeling (see Sect. 3.2). It should be noticed that here the value of D is fixed to $D = 5$ h to determine the neighborhoods. The Table 1 summarizes the performances of each algorithm to compare the PSO performance to the exact solution provided by the MILP model on the 32 scenarios of Σ. Here the time limit of GUROBI^{TM}was fixed to 1 h and therefore the solution is not displayed when the time has been exceeded. It should be also noticed that each instance is solved with $m = 20$ particles and $l = 1000$ cycles. We may easily observe that in every tested scenario, PSO_2 provides better or equal results than PSO_1. Furthermore, we may identify several scenarios with which the two PSO algorithms provide these optimal results. These particular cases correspond to simple schedules and create the tapered shapes of the error ratio. In deed, it appears than for instance most of the time schedule 15 surgical operations on a working weak (5 days) implies to allocate 3 operations to each day while schedule 16 surgical operations is not a trivial problem.

6   Conclusion

The current economic context and the medical sector stakes are conductive to hospital groups creation. It is clear that these new structures imply rethinking the hospital context logistics in its entirety. The synchronization between the operating theater and the sterilization service of the pharmacy constitutes a large operating research field full of possibilities that must be explored. Indeed, in a hospital group with central pharmacy, the latter became a major issue of the complete functioning and must be considered as such. With the study presented in this paper, we hope to lay the foundations of a hospital transverse logistic taking into account the issues of these two main structures. The purpose here is to improve the surgical devices flow which are circulated through the hospital group. The next step is to provide a complete planning creation method by improving several aspects of the detailed process. First, more operating theater aspects must be considered by adding new constraints and then modifying the objective function. We contemplate using multi-objective function or Pareto frontier, the advantage of the latter could be to compute a set of schedules and let the final decision to the hospital group management department. Moreover, the presented process must be adapted for real time environment. Indeed, in this paper we detailed a scheduling problem considering that every surgical procedure is already known but in real case the procedures are prescribed time after time by the surgeons. We may imagine for instance to schedule the operation at the end of each day considering the ones already planed.

Several aspects of the pharmacy sector can still be studied to improve the entire hospital complexes functioning as medicines ordering, inter sites transportation, stock managing and so on. By considering more common aspects of the pharmacy and the hospital we may include each logistic decision into a complete logistic covering the entire hospital group context.

References

1. Beroule, B., Grunder, O., Barakat, O., Aujoulat, O., Lustig, H.: Ordonnancement des interventions chirurgicales d'un hopital avec prise en compte de l'étape de stérilisation dans un contexte multi-sites (2016)
2. Cardoen, B., Demeulemeester, E., Beliën, J.: Optimizing a multiple objective surgical case sequencing problem. Int. J. Prod. Econ. 119(2), 354–366 (2009)
3. Cardoen, B., Demeulemeester, E., Beliën, J.: Sequencing surgical cases in a daycare environment: an exact branch-and-price approach. Comput. Oper. Res. 36(9), 2660–2669 (2009)
4. Cardoen, B., Demeulemeester, E., Beliën, J.: Operating room planning and scheduling: a literature review. Eur. J. Oper. Res. 201(3), 921–932 (2010)
5. Clerc, M., Kennedy, J.: The particle swarm-explosion, stability, and convergence in a multidimensional complex space. IEEE Trans. Evol. Comput. 6(1), 58–73 (2002)
6. Clerc, M., Siarry, P.: Une nouvelle métaheuristique pour l'optimisation difficile: la méthode des essaims particulaires. J3eA, 3:007 (2004)
7. Dexter, F., Blake, J.T., Penning, D.H., Sloan, B., Chung, P., Lubarsky, D.A.: Use of linear programming to estimate impact of changes in a hospital's operating room time allocation on perioperative variable costs. Anesthesiology 96(3), 718–724 (2002)
8. Dexter, F., Epstein, R.H.: Scheduling of cases in an ambulatory center. Anesthesiol. Clin. North Am. 21(2), 387–402 (2003)
9. Dexter, F., Macario, A.: When to release allocated operating room time to increase operating room efficiency. Anesth. Analg. 98(3), 758–762 (2004)
10. Dexter, F., Macario, A., Traub, R.D.: Which algorithm for scheduling add-on elective cases maximizes operating room utilization? Use of bin packing algorithms, fuzzy constraints in operating room management. Anesthesiology 91(5), 1491–1500 (1999)
11. Fei, H., Meskens, N., Chu, C.: A planning and scheduling problem for an operating theatre using an open scheduling strategy. Comput. Ind. Eng. 58(2), 221–230 (2010)
12. Graham, R.L., Lawler, E.L., Lenstra, J.K., Kan, A.R.: Optimization and approximation in deterministic sequencing and scheduling: a survey. Ann. Discrete Math. 5, 287–326 (1979)
13. Guinet, A., Chaabane, S.: Operating theatre planning. Int. J. Prod. Econ. 85(1), 69–81 (2003)

14. Hanset, A., Fei, H., Roux, O., Duvivier, D., Meskens, N.: Ordonnancement des interventions chirurgicales par une recherche tabou: Exécutions courtes vs longues. Logistique et Transport LT 2007 (2007)
15. Jebali, A., Alouane, A.B.H., Ladet, P.: Operating rooms scheduling. Int. J. Prod. Econ. **99**(1), 52–62 (2006)
16. Kenndy, J., Eberhart, R.C.: Particle swarm optimization. In: Proceedings of IEEE International Conference on Neural Networks, vol. 4, pp. 1942–1948 (1995)
17. Kennedy, J.: Particle swarm optimization. In: Sammut, C., Webb, G.I. (eds.) Encyclopedia of Machine Learning, pp. 760–766. Springer, New York (2011)
18. Klement, N.: Planification et affectation de ressources dans les réseaux de soin: analogie avec le problème du bin packing, proposition de méthodes approchées. Ph.D. thesis, Université Blaise Pascal-Clermont-Ferrand II (2014)
19. Marcon, E., Dexter, F.: Impact of surgical sequencing on post anesthesia care unit staffing. Health Care Manage. Sci. **9**(1), 87–98 (2006)
20. Omran, M.G.H., Salman, A., Engelbrecht, A.P.: Dynamic clustering using particle swarm optimization with application in image segmentation. Pattern Anal. Appl. **8**(4), 332–344 (2006)
21. Pandey, S., Wu, L., Guru, S.M., Buyya, R.: A particle swarm optimization-based heuristic for scheduling workflow applications in cloud computing environments. In: 2010 24th IEEE International Conference on Advanced Information Networking and Applications, pp. 400–407. IEEE (2010)
22. Robinson, J., Rahmat-Samii, Y.: Particle swarm optimization in electromagnetics. IEEE Trans. Antennas Propag. **52**(2), 397–407 (2004)
23. Saadani, N.H., Guinet, A., Chaabane, S.: Ordonnancement des blocs operatoires. In: MOSIM: Conference francophone de MOdélisation et SIMulation, vol. 6 (2006)
24. Shi, Y., Eberhart, R.: A modified particle swarm optimizer. In: The 1998 IEEE International Conference on Evolutionary Computation Proceedings, IEEE World Congress on Computational Intelligence, pp. 69–73. IEEE (1998)
25. Shi, Y., Eberhart, R.C.: Parameter selection in particle swarm optimization. In: Porto, V.W., Saravanan, N., Waagen, D., Eiben, A.E. (eds.) EP 1998. LNCS, vol. 1447, pp. 591–600. Springer, Heidelberg (1998). doi:10.1007/BFb0040810
26. Trelea, I.C.: The particle swarm optimization algorithm: convergence analysis and parameter selection. Inf. Process. Lett. **85**(6), 317–325 (2003)
27. Wullink, G., Van Houdenhoven, M., Hans, E.W., van Oostrum, J.M., van der Lans, M., Kazemier, G.: Closing emergency operating rooms improves efficiency. J. Med. Syst. **31**(6), 543–546 (2007)

Data Exchange Topologies for the DISCO-HITS Algorithm to Solve the QAP

Omar Abdelkafi$^{(\boxtimes)}$, Lhassane Idoumghar, Julien Lepagnot,
and Mathieu Brévilliers

Université de Haute-Alsace (UHA), LMIA (E.A. 3993),
4 Rue des Frères Lumière, 68093 Mulhouse, France
{omar.abdelkafi,lhassane.idoumghar,julien.lepagnot,
mathieu.Brevilliers}@uha.fr

Abstract. Exchanging information between processes in a distributed environment can be a powerful mechanism to improve results for combinatorial problem. In this study, we propose three exchange topologies for the distance cooperation hybrid iterative tabu search algorithm called DISCO-HITS. These topologies are experimented on the quadratic assignment problem. A comparison between the three topologies is performed using 21 well known instances of size between 40 and 150. Our algorithm produces competitive results and can outperform algorithms from the literature for many benchmark instances.

Keywords: Metaheuristics · DISCO-HITS · Quadratic assignment problem · Topologies

1 Introduction

The Quadratic assignment problem (QAP) is an NP-hard problem. It is well known for its multiple applications. Many practical problems in electronic, chemistry, transport, industry and many others can be formulated as QAP. This problem was first introduced by Koopmans and Beckmann [1] to model a facility location problem. It can be described as the problem of assigning a set of facilities to a set of locations with given distance and flow between locations and facilities, respectively. The objective is to place the facilities on locations in such a way that the sum of the products between flows and distances is minimized. The problem can be formulated as follows:

$$\min_{p \in P} z(p) = \sum_{i=1}^{n} \sum_{j=1}^{n} f_{ij} d_{p(i)p(j)} \tag{1}$$

where f and d are the flow and distance matrices respectively, $p \in P$ represents a solution where p_i is the location assigned to facility i and P is the set of all n vector permutations. The objective is to minimize $z(p)$, which is the total cost assignment for the permutation p.

© Springer International Publishing AG 2016
P. Siarry et al. (Eds.): ICSIBO 2016, LNCS 10103, pp. 57–64, 2016.
DOI: 10.1007/978-3-319-50307-3_4

In this work, we propose an experimental analysis of different exchanging topologies to solve the QAP. The aim of this work is to explore the influence of these topologies. The parallel level used is the algorithmic level [2].

The rest of the paper is organized as follows. In Sect. 2, we review some of the best-known distributed approaches to solve the QAP. In Sect. 3, we describe the different topologies used in this work. Section 4 shows the experimental results for a set of QAPLIB instances. Finally, in Sect. 5, we conclude the paper and we propose some perspectives.

2   Background

Since its introduction in 1957 [1], the QAP became an important problem in theory and practice. It can be considered as one of the hardest combinatorial problems due to its computational complexity. Different metaheuristics have been proposed to provide competitive results [3–7].

The parallel and distributed design of metaheuristic approaches has the capacity to improve the solution quality and to reduce the execution time. The computational cost of the QAP and its difficult search space make this problem suitable for parallelization. The parallel and distributed design of metaheuristics to solve the QAP is underexploited. Very few works propose it, such as the Robust Tabu search (Ro-Ts) [3] which is a parallelization of neighborhood between different processors.

In 2001, a parallel model of ant colonies is proposed [8]. A central memory to manage all communications of the search information is implemented in the master process. The search information is composed of the pheromone matrix and the best solution found. At each iteration, the master broadcasts the pheromone matrix to all the ants. Each process represents one ant and each ant constructs a complete solution and applies a Tabu Search (TS) in parallel. The process sends the solution found and the local pheromone matrix to the master. The master updates the search information. In 2005, a parallel path-relinking algorithm is proposed [9]. This proposition generates different solutions by applying path-relinking to a set of trial solutions. To improve the solutions created by the path-relinking procedure, the Ro-Ts algorithm is run in parallel starting with different trial solutions. It allows the reduction of the execution time but it does not change the behavior of the sequential algorithm and the solution quality. In 2009, a cooperative parallel TS algorithm for the QAP is introduced [6]. This approach initializes as many starting solutions as there are available processors. Each processor executes one independent TS in parallel. The initialization phase provides good starting solutions while maintaining some level of diversity. After the initialization, at each iteration, all the processors execute a TS in parallel. At the end of the generation, the current processor compares its solution with its neighbor process. If the neighbor process gets better results, the current process replaces its current solution with a mutated copy of the neighbor solution. In 2015, a parallel hybrid algorithm is proposed [10]. This proposition is composed of three steps. The first step is the seed generation which consists in using a parallel Genetic Algorithm (GA) based on the island model. Each process represents an island and at each generation, the master broadcasts the global best solution to all islands. All nodes execute a GA in parallel. The second step is the TS diversification. This method is applied to all the parallel nodes. Finally, the global best solution obtained with the first two steps is used as an initial seed for the Ro-Ts.

3 Topologies to Exchange Information Between Processes

In 2015, a cooperative Iterative Tabu Search (ITS) called DIStance COoperation between Hybrid Iterative Tabu Search (DISCO-HITS) is proposed [11]. Each process performs an ITS in which a Ro-Ts is executed at each generation. After each iteration, each process sends its current solution to the neighbor process. Then, a distance is computed between the current solution and the solution received from the neighbor process. According to this distance, the algorithm takes the decision to apply the uniform crossover (UX), to perturb the solution or to make a re-localization of this solution. Algorithm 1 presents the DISCO-HITS version used in this paper.

Algorithm 1. Distance Cooperation Between Hybrid Iterative Tabu Search

1: **Input:** *perturb:* % perturbation; *n:* size of solution; *cost:* cost of the current solution; *Fcost:* best cost found; $S_{current}$: current solution; S_{best}: best solution found; S_{EX}: solution exchanged;
2: Initialization of the solution for the current process;
3: **repeat**
4:    TS algorithm; [3]
5:    **if** cost < Fcost **then**
6:       $Fcost = cost$;
7:       Update the S_{best} with $S_{current}$;
8:    **end if**
9:    level = 0; counter = 0;
10:   Exchange $S_{current}$ between processes;
11:   **for** i = 0 to n /* Compute distances */ **do**
12:      **if** $S_{current}[i] == S_{EX}[i]$ **then**
13:         counter ++;
14:      **end if**
15:   **end for**
16:   **if** counter < $\frac{n}{4}$ **then**
17:      level = 0; /* Big distance between the two processes */
18:   **else**
19:      **if** counter < $\frac{3 \times n}{4}$ **then**
20:         level = 1; /* Processes are relatively close */
21:      **else**
22:         level = 2; /* Processes are very close */
23:      **end if**
24:   **end if**
25:   **if** level == 0 **then**
26:      Update $S_{current}$ with the **UX** of S_{best};
27:   **else**
28:      **if** level == 1 **then**
29:         **Perturbation** of $S_{current}$ with the **perturb** parameter;
30:      **else**
31:         **Re-localization** of $S_{current}$;
32:      **end if**
33:   **end if**
34: **until** (Stop condition)

Exchanging information between processes (Algorithm 1 line 10) is performed according to a topology. Algorithm 1 sends its current solution to one process and receives the current solution of another process. We propose three topologies in this paper. All the topologies are defined with a sequence. Process with index i sends to process with index $i+1$ and receives from index $i-1$. The last index sends its information for the first index to close the circle of exchange. This method ensures the sending and receipt of only one solution.

The first topology is the classical ring architecture implemented in the variant called *DISCO-RING-UX*. Each process sends its current solution to the next process and receives from the previous process. For example, if we use four processes, the sequence of exchange is $\{0; 1; 2; 3\}$. with this sequence, process 2 sends to process 3 and process 3 sends to process 0. This sequence is constant from the beginning of the execution to the end. The aim of this topology is to experiment a constant impact between two processes.

The second topology is the random architecture implemented in the variant called *DISCO-RANDOM-UX*. Each process sends its current solution to a random process and receives from a random process. For example if we use four processes the sequence of exchange can be $\{1; 2; 0; 3\}$. This sequence is randomly perturbed before each exchange. The aim of this topology is to experiment a dynamic impact between two processes. The random exchange allows a better diversification.

The last topology is a learning sequence architecture based on the fast ant algorithm implemented in the variant called *DISCO-LEARNING-UX*. In this case, our ant is the sequence of exchange. If the previous sequence allows the algorithm to improve, a quantity of pheromone is deposited for the pair of processes which exchange the current solution. Otherwise, the quantity of pheromone deposited is significantly reduced. Before the exchanging step, the pheromone matrix is updated and the ant is reconstructed. After the reconstruction, a step of evaporation is performed. The aim is to learn the best topology to exchange information by converging to the best sequence.

4   Experimental Results

4.1   Platform and Tests

In our experimentation, the algorithm is written in C/C++. It runs on a cluster of 8 machines Intel Core processor i5-3330 CPU (3.00GHz) with 4 GB of RAM and an NVIDIA GeForce GTX680 GPU. The proposed algorithm is experimented on benchmark instances from the QAPLIB [13]. The size of the instances varies between 40 and 150. Every instance is executed 10 times and the average results of these executions are given in the experiments. All the results are expressed as a percentage deviation from the best known solutions (BKS) (Eq. 2).

$$deviation = \frac{(solution - BKS) \times 100}{BKS} \qquad (2)$$

Table 1. Parameter of *DISCO-HITS*

Parameters	Value
TSiteration	$1000 \times n$
Global iteration	200
Aspiration criteria	$n \times n \times 5$
Percentage of perturbation	25 %

Table 2. Comparison of different topologies

Instance(21)	BKS	DISCO-RING-UX		DISCO-RANDOM-UX		DISCO-LEARNING-UX	
		Deviation	Time	Deviation	Time	Deviation	Time
tai40a	3139370	0.067(1)	3.59	**0.059(2)**	3.4	0.067(1)	3.6
tai50a	4938796	0.317(0)	6.65	0.344(0)	6.6	**0.308(0)**	6.7
tai60a	7205962	0.401(0)	11.6	0.400(0)	11.4	**0.317(0)**	11.4
tai80a	13515450	0.605(0)	27.2	0.613(0)	27.1	**0.590(0)**	27.2
tai100a	21052466	0.493(0)	53.9	0.478(0)	53.8	**0.462(0)**	53.8
tai50b	458821517	0.000(10)	6.5	0.000(10)	6.5	0.000(10)	6.6
tai60b	608215054	0.000(10)	11.3	0.000(10)	11.2	0.000(10)	11.3
tai80b	818415043	0.000(10)	27	0.000(10)	26.9	0.000(10)	27
tai100b	1185996137	0.000(10)	53.2	0.000(10)	53	0.000(10)	53.2
tai150b	498896643	0.151(0)	190	**0.129(0)**	189	0.139(0)	196.1
sko72	66256	0.001(8)	19.6	**0.000(10)**	19.5	0.001(9)	19.7
sko81	90998	0.004(6)	28	0.004(6)	28	**0.002(8)**	28.1
sko90	115534	0.001(8)	38.5	**0.000(10)**	38.6	0.001(8)	38.6
sko100a	152002	0.005(6)	53.5	**0.004(8)**	53.5	0.005(8)	53.5
sko100b	153890	0.002(8)	53.5	**0.001(9)**	53.3	0.002(8)	53.5
sko100c	147862	0.002(1)	53.5	**0.001(6)**	53.3	0.001(2)	53.5
sko100d	149576	0.004(4)	53.5	**0.002(5)**	53.4	0.005(4)	53.5
sko100e	149150	0.002(6)	53.7	**0.002(8)**	53.3	0.002(7)	53.4
sko100f	149036	0.004(3)	53.6	0.006(3)	53.8	**0.003(4)**	53.4
wil100	273038	0.003(1)	53.6	0.003(2)	53.5	**0.002(3)**	53.6
tho150	8133398	**0.016(0)**	198.1	0.030(0)	189.3	0.021(0)	191.4
Average type 2		0.3766(1)	20.6	0.3788(2)	20.5	**0.3488(1)**	20.5
Average type 3		0.0302(40)	57.6	**0.0258(40)**	57.3	0.0278(40)	58.8
Average type 4		**0.0040(51)**	59.9	0.0048(67)	59	0.0041(61)	59.3
Average		0.099(92)	50	0.099(109)	49.5	**0.092(102)**	49.9

The QAPLIB archive comprises 136 instances that can be classified into four types: real life instances (type 1); unstructured randomly generated instances based on a uniform distribution (type 2); randomly generated instances similar to real life instances (type 3); instances in which distances are based on the Manhattan distance on a grid (type 4).

Table 3. Comparison with the literature

Instance(19)	BKS	DISCO-RING-UX		DISCO-RANDOM-UX		DISCO-LEARNING-UX		TLBO-RTS		CPTS	
		Deviation	Time	Deviation	Time	Deviation	Time	Deviation	Time	Deviation	Time
tai40a	3139370	0.067(1)	3.59	0.059(2)	3.4	0.067(1)	3.6	**0.000**	29	0.148(1)	3.5
tai50a	4938796	0.317(0)	6.65	0.344(0)	6.6	**0.308(0)**	6.7	0.360	55	0.440(0)	10.3
tai60a	7205962	0.401(0)	11.6	0.400(0)	11.4	**0.317(0)**	11.4	0.410	95.3	0.476(0)	26.4
tai80a	13515450	0.605(0)	27.2	0.613(0)	27.1	**0.590(0)**	27.2	0.870	239.5	0.691(0)	94.8
tai100a	21052466	0.493(0)	53.9	0.478(0)	53.8	**0.462(0)**	53.8	0.596	483.3	0.589(0)	261.2
tai80b	818415043	0.000(10)	27	0.000(10)	26.9	0.000(10)	27	0.000	239	0.000(10)	110.9
tai100b	1185996137	0.000(10)	53.2	0.000(10)	53	0.000(10)	53.2	0.000	508.2	0.001(8)	241
tai150b	498896643	0.151(0)	190	0.129(0)	189	0.139(0)	196.1	**0.015**	428.5	0.076(0)	7377.8
sko72	66256	0.001(8)	19.6	0.000(10)	19.5	0.001(9)	19.7	0.000	172.8	0.000(10)	69.6
sko81	90998	0.004(6)	28	0.004(6)	28	0.002(8)	28.1	0.000	348.2	0.000(10)	121.4
sko90	115534	0.001(8)	38.5	0.000(10)	38.6	0.001(8)	38.6	0.000	342.8	0.000(10)	193.7
sko100a	152002	0.005(6)	53.5	0.004(8)	53.5	0.005(8)	53.5	0.003	594.3	**0.000(10)**	304.8
sko100b	153890	0.002(8)	53.5	0.001(9)	53.3	0.002(8)	53.5	0.005	482.6	**0.000(10)**	309.6
sko100c	147862	0.002(1)	53.5	0.001(6)	53.3	0.001(2)	53.5	0.000	508.5	0.000(10)	316.1
sko100d	149576	0.004(4)	53.5	0.002(5)	53.4	0.005(4)	53.5	0.009	509.4	**0.000(10)**	309.8
sko100e	149150	0.002(6)	53.7	0.002(8)	53.3	0.002(7)	53.4	0.005	614.5	**0.000(10)**	309.1
sko100f	149036	0.004(3)	53.6	0.006(3)	53.8	0.003(4)	53.4	0.005	482.6	0.003(4)	310.3
wil100	273038	0.003(1)	53.7	0.003(2)	53.5	0.002(3)	53.6	0.000	482.6	0.000(10)	316.6
tho150	8133398	0.016(0)	198.1	0.030(0)	189.3	0.021(0)	191.4	0.030	556.6	**0.013(0)**	1991.7
Average type 2		0.3766(1)	20.6	0.3788(2)	20.5	**0.3488(1)**	20.5	0.4472	180.42	0.4688(1)	79.2
Average type 3		0.0503(20)	90	0.0430(20)	89.6	0.0463(20)	92.1	**0.0050**	391.9	0.0257(18)	2576.6
Average type 4		0.0040(51)	59.9	0.0048(67)	59	0.0041(61)	59.3	0.0052	463.2	**0.0014(94)**	413.9
Average		0.109(72)	54.3	0.109(89)	53.7	**0.101(82)**	54.3	0.121	377.5	0.128(113)	667.3
Average NOFE		1.48e+08		1.48e+08		1.48e+08		7.55e+10		9.23e+08	

4.2   Parameters

DISCO-HITS contains a set of parameters. A set of experimentation is executed to fix all the parameters. Table 1 shows the parameters used in the experimentation, where n is the size of the problem and *rank* is the index of the current process.

4.3   Experimentation of the Three Topologies

Table 2 contains the results for the three variants proposed in this work. The same number of objective function evaluations and the same machines are used (equivalent computing power). The time is expressed in minutes. The number within brackets is the number of times each algorithm gets the BKS among the 10 trials.

Through the 21 benchmark instances presented in this work, *DISCO-RING-UX* outperforms all the variants for only one instance (tho150 in type 2). *DISCO-RANDOM-UX* outperforms all the variants for 9 instances especially from type 4. Finally, *DISCO-LERNING-UX* outperforms all the variants for 7 instances especially from type 3. *DISCO-LERNING-UX* gets the best global average of 0.092 %. This variant shows the most stable results for the 3 types.

4.4   Literature Comparison

Table 3 presents several comparisons with two distributed algorithms from the literature. Cooperative parallel tabu search (CPTS) [6] (2009) and Teaching-Learning-Based Optimization (TLBO) [12] (2015).

The average number of objective function evaluation (NOFE in Table 3) used in our 3 variants is much lower than for the literature algorithms. *CPTS* algorithm uses 5.8 times more objective function evaluations and *TLBO* uses 523.5 times more evaluations. We use 19 well-known benchmark instances from the QAPLIB which are difficult to solve. *DISCO-LERNING-UX* outperforms all the algorithms on 4 instances from type 3. *TLBO* outperforms all the algorithms on 2 instances (tai40a and tai150b). *CPTS* outperforms all the algorithms on 5 instances from type 4. *DISCO-LERNING-UX* gets the best global average of 0.101 % against 0.128 % for *CPTS* and 0.121 % for *TLBO*. Considering the difference of NOFE, the results obtained by our 3 variants are very competitive.

5   Conclusion and Perspectives

In this work, we have presented and experimented three variants of the DISCO-HITS algorithm with different topologies to solve the QAP. The results show that the proposed variants perform efficiently. We evaluated our variants on 19 benchmark instances from the QAPLIB and they get the best average results compared to two leading distributed algorithms from the literature.

In summary, the main contributions of this work are the proposition of these variants and the experimentation of three different topologies to exchange information in a distributed environment. The automatically learnt topology, used in the *DISCO-LERNING-UX* variant, shows the best average results.

In future works, there are several possible ways to extend this work. One possibility is to experiment other parameters to get better results on large neighborhood instances. An experimental analysis can also be made using some instances which are not explored in literature, such as tai729eyy. Finally, this approach can be experimented for other combinatorial problems to analyze its behavior with other kinds of problems.

References

1. Koopmans, T., Beckmann, M.: Assignment problems and the location of economic activities. Econometrica **25**(1), 53–76 (1957)
2. Talbi, E.G.: Metaheuristics: From Design to Implementation. University of Lille - CNRS - INRIA, John wiley and sons Inc. (2009)
3. Taillard, E.: Robust taboo search for the quadratic assignement problem. Parallel Comput. **17**, 443–455 (1991)
4. James, T., Rego, C., Glover, F.: Multistart tabu search and diversification strategies for the quadratic assignment problem. IEEE Trans. Syst. Man Cybern. Part A Syst. Hum. **39** (3), 579–596 (2009)
5. Benlic, U., Hao, J.K.: Breakout local search for the quadratic assignement problem. Appl. Math. Comput. **219**, 4800–4815 (2013)
6. James, T., Rego, C., Glover, F.: A cooperative parallel tabu search algorithm for the quadratic assignment problem. Eur. J. Oper. Res. **195**, 810–826 (2009)
7. Czapinski, M.: An effective parallel multistart tabu search for quadratic assignment problem on CUDA platform. J. Parallel Distrib. Comput. **73**, 1461–1468 (2013)
8. Talbi, E.G., Roux, O., Fonlupt, C., Robillard, D.: Parallel ant colonies for the quadratic assignment problem. Future Gener. Comput. Syst. **17**, 441–449 (2001)
9. James, T., Rego, C., Glover, F.: Sequential and parallel path relinking algorithms for the quadratic assignment problem. IEEE Intell. Syst. **20**(4), 58–65 (2005)
10. Tosun, U.: On the performance of parallel hybrid algorithms for the solution of the quadratic assignment problem. Eng. Appl. Artif. Intell. **39**, 267–278 (2015)
11. Abdelkafi, O., Idoumghar, L., Lepagnot, J.: Comparison of two diversification methods to solve the quadratic assignment problem. Procedia Comput. Sci. **51**, 2703–2707 (2015)
12. Dokeroglu, T.: Hybrid teaching-learning-based optimization algorithms for the quadratic assignment problem. Comput. Ind. Eng. **85**, 86–101 (2015)
13. Burkard, R.E., Karisch, S.E., Rendl, F.: QAPLIB - a quadratic assignment problem library. J. Glob. Optim. **10**(4), 391–403 (1997)

Distributed Local Search for Elastic Image Matching

Hongjian Wang$^{(\boxtimes)}$, Abdelkhalek Mansouri, Jean-Charles Créput,
and Yassine Ruichek

IRTES-SeT, Université de Technologie de Belfort-Montbéliard,
90010 Belfort, France
hongjian3715@gmail.com

Abstract. We propose a *distributed local search* (DLS) algorithm, which
is a parallel formulation of a local search procedure in an attempt to follow the spirit of standard local search metaheuristics. Applications of different operators for solution diversification are possible in a similar way
to *variable neighborhood search*. We formulate a general energy function
to be equivalent to elastic image matching problems. A specific example
application is *stereo matching*. Experimental results show that the GPU
implementation of DLS seems to be the only method that provides an
increasing acceleration factor as the instance size augments, among eight
tested energy minimization algorithms.

Keywords: Parallel and distributed computing · Variable neighborhood search · Stereo matching · Graphics processing unit

1 Introduction

Local search, also referred as hill climbing, descent, iterative improvement, general single-solution based metaheuristics and so on, is a metaheuristic algorithm [1]. Starting with a given initial solution, at each iteration the heuristic
replaces the current solution by a neighbor solution that improves the fitness
function. The search stops when all candidate neighbors are worse than the current solution, meaning a local optimum is reached. Existing parallelization strategies for local search can be divided into three categories. In the first category, the
evaluation of neighborhood is made in parallel [2,3]; in the second category, the
focus is on the parallel evaluation of a single solution, and the function can be
viewed as an aggregation of partial functions [2,4]; in the third category, several
local search metaheuristics are simultaneously launched for computing robust
solutions [5,6]. In our opinion, an interesting parallel implementation model of
local search should be fully distributed, where each processor carries out its own
neighborhood search based on some parts of the input data, considering only a
local part of the whole solution. Operations on different processors should be similar, with no centralized selection procedure, except for final evaluation. A final
solution should be obtained with the partial operations from different processors.

© Springer International Publishing AG 2016
P. Siarry et al. (Eds.): ICSIBO 2016, LNCS 10103, pp. 65–74, 2016.
DOI: 10.1007/978-3-319-50307-3_5

Following this idea, we propose a *distributed local search* (DLS) algorithm and implement it on GPU parallel computing platforms.

A natural field of applications with GPU processing is image processing, which is a domain at the origin of GPU development. A lot of image processing and computer vision problems can be viewed as optimization problems in a more general way, dealing with brute data distributed in some Euclidean space and system in relation to the data. More often, these NP-hard optimization problems involve data distributed in the plane and elastic structures represented by graphs that must match the data. Such optimization problems can be stated in a generic framework of *graph matching* [7,8]. In this paper, we are particularly interested in moving grids in the plane following the idea of visual correspondence problem, which is to compute the pairs of pixels from two images that result from the same scene element. A typical example application is *stereo matching*, which we formulate as an elastic image matching problem [9]. We apply the proposed DLS algorithm to stereo matching by minimizing the corresponding energy function.

The DLS can be used for parallel implementation of elastic matching problems that include not only visual correspondence problems but also neural network topological maps, or elastic nets approaches [10,11], modeling the behavior of interacting components inspired by biological systems and collective behaviors at a low level of granularity. The framework is based on data decomposition, with the idea of modeling the geometry of objects using some adaptive (elastic) structures that move in space and continuously interact with the input data distribution memorized into a cellular matrix [12]. Then spatial metaphors, as well as biological metaphors should fit well into the cellular matrix framework.

The rest of this paper is organized as follows. In Sect. 2, we formulate a general energy function to be equivalent to elastic image matching problems. In Sect. 3, we present the DLS algorithm in detail, providing basic data structures and operations in Subsect. 3.1, explaining local evaluation in Subsect. 3.2, designing two classes of move operators in Subsect. 3.3, and giving the details of GPU implementation in Subsect. 3.4. Experimental results are reported in Sect. 4, before some conclusions are drawn in Sect. 5.

2   Elastic Grid Matching

We define a class of visual correspondence problems as *elastic grid matching* problems, where we use a two-dimensional grid to represent an image. Given two input grids with same size and same regular topology, one is a matcher grid $G_1 = (V_1, E_1)$ where a vertex is a pixel (from the corresponding image) with a variable location in the plane, while the other is a matched grid $G_2 = (V_2, E_2)$ where vertices are pixels located in a regular grid. The goal of elastic grid matching is to find the matcher vertex locations in the plane, so that the following energy function

$$E(G_1) = \sum_{p \in V_1} D_p(p - p_0) + \lambda \cdot \sum_{\{p,q\} \in E_1} V_{p,q}(p - p_0, q - q_0) \tag{1}$$

is minimized, where p_0 and q_0 are the default locations of p and q respectively in a regular grid. Here, D_p is the data energy that measures how much assigning label f_p to pixel p disagrees with the data, and $V_{p,q}$ is the smoothness energy that expresses smoothness constraints on the labelings enforcing spatial coherence [13–15]. A label f_p in visual correspondence represents a pixel moving from its regular position into the direction of its homologous pixel, $i.e.$ $f_p = p - p_0$. In the following sections, we will directly use the notations of labels as relative displacements, as usual with such problems. The energy function is commonly used for visual correspondence problems, and it can be justified in terms of maximum a posteriori estimation of a *Markov random field* (MRF) [16,17].

It has been proven that elastic image matching is NP-complete [9], and finding the global minimum for the energy function even with the simplest smoothness penalty, the piecewise constant prior, is NP-hard [13,14]. We choose the local search metaheuristics to deal with the energy minimization problem.

3   Distributed Local Search

Based on the cellular matrix model proposed in [12], we design a parallel local search algorithm, the DLS, to implement many local search operations on different parts of the data in a distributed way. It is a parallel formulation of local search procedures in an attempt to follow the spirit of standard local search metaheuristics. Starting from its location in the cellular matrix, each processor locally acts on the data located in the corresponding cell according to the cellular decomposition, in order to achieve local evaluation, perform neighborhood search, and select local improvement moves to execute. The many processes locally interact in the plane, making evolve the current solution into an improved one. The solution results from the many independent local search operations simultaneously performed on the distributed data in the plane. Normally, a local search algorithm with single operator obtains local minima. In order to escape from local minima, we design several operators. Applications of different operators for diversification are possible in a similar way to the *variable neighborhood search* (VNS).

3.1   Data Structures and Basic Operations

The data structures and direction of operations for DLS algorithms are illustrated in Fig. 1. The input data set is deployed on the low level of both matcher grid and matched grid, represented as regular images in the figure. The honeycomb cells represent the cellular matrix level of operations. Each cell is a basic processor that handles a basic local search processing iteration with the three following steps: neighborhood generation (**get**); neighbor solution evaluation and selecting the best neighbor (**search**); then moving the matcher grid toward the selected neighbor solution (**operate**). The nature and size of specific moves and neighborhoods will depend on the type of used operator and the level of cellular matrix. The higher is the level, the larger is the local cell/neighborhood. In the

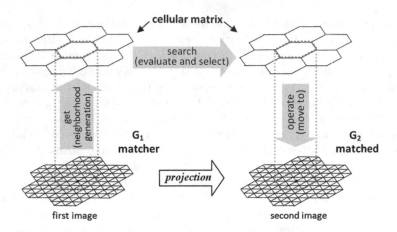

Fig. 1. Basic projection for DLS.

cellular matrix model, a solution is composed of many sub-solutions from many cells. Each sub-solution is evolved from an initial sub-solution based on the distributed data in a cell. By partitioning the data and solution, the neighborhood structure is also partitioned at the same time.

3.2 Local Evaluation with Mutual Exclusion

During the parallel operation, the coherence of local evaluation with mutual exclusion is violated by conflict operations. A conflict operation occurs when a same pixel or two neighboring pixels is/are being evaluated and moved simultaneously by two threads. Conflict operations only happen on frontier pixels, which are the pixels on the cell frontiers according to the cellular matrix partition of the image. In order to eliminate the conflict operations in DLS, we propose a strategy, called *dynamic change of cell frontiers* (DCCF), by which we limit the move to the internal pixels of a cell only. Cell frontier pixels remain at fixed locations, and they are not concerned by local moves so that exclusive access of the thread to its internal region delimited by the cell is guaranteed. A problem that arises is how to manage cell frontier pixels and make them participate in the optimization process. As a solution, the cellular matrix decomposition is dynamically changeable from the CPU side before the application of a round of DLS operations. At different moments, the cellular matrix decomposition slightly shifts on the input image in order to change the cell frontiers and consequently the fixed pixels. For a given cellular matrix decomposition, cell frontier pixels are then fixed and not allowed to be moved by current DLS operations.

3.3 Neighborhood Operators

We design different neighborhood operators for the DLS algorithm applied to elastic grid matching. We use the notations of labeling problems to present these operators. Move operations in a given neighborhood structure correspond to changing labels of pixels in the corresponding labeling space. Operators are classified between small moves and large moves. In the first case, only a single pixel from a given cell moves at a time; in the second case, larger sets of pixels from a given cell can simultaneously move.

Small move operators. In a move operation, if only one pixel moves, meaning that only one pixel's label is changed, this kind of operation is called small move operation. We design two small move operators: *local move operator* that applies an increment/decrement to the current label of the considered pixel; *propagation operator* that takes the labels of the considered pixel's neighboring pixels, as candidate labels, and then replaces the current label with the best one found in a propagation window.

Large move operators. They consider multiple pixels. We design six large move operators: *random pixels move operator* randomly picks several pixels in the considered cell, and then assigns a same candidate label to these pixels; *random pixels jump operator* randomly picks several pixels in the considered cell, and then applies a same increment/decrement to the current labels of the considered pixels; *random pixels expansion operator* randomly picks two groups of pixels, where pixels in the same group have the same label, and then "expands" the label of one group to the other, setting the labels of all the pixels in the second group with the same label as the first group; *random pixels swap operator* picks pixels in the same way as the random pixels expansion operator does, and then "swaps" the labels of the two groups, setting the labels of all the pixels in the second group with the label of the first group, meanwhile setting the labels of all the pixels in the first group with the label of the second group; *random window move operator* picks a fixed-sized window of pixels at a random position within the considered cell, and then assigns a same candidate label to all the pixels in this picked window; *random window jump operator* picks pixels in the same way as the random window move operator does, and then applies a same increment/decrement to the current labels of all the pixels in this picked window. More details about these operators can be found in [12].

3.4   GPU Implementation Under VNS Framework

We implement the DLS algorithm on GPU platforms in Compute Unified Device Architecture (CUDA). The CUDA kernel calling sequence from the CPU side enables the application of different operators in the spirit of VNS and manages dynamic changes of cellular matrix frontiers. According to our previous experiments, the repartition of tasks between host (CPU) and device (GPU) is actually the best compromise we found to exploit the GPU CUDA platform at a reasonable level of computation granularity.

The flow chart executed from CPU side is presented in Fig. 2. The data transfer between CPU side and GPU side only occurs at the beginning and the end of the algorithm. The two kernels that are called from CPU side and executed

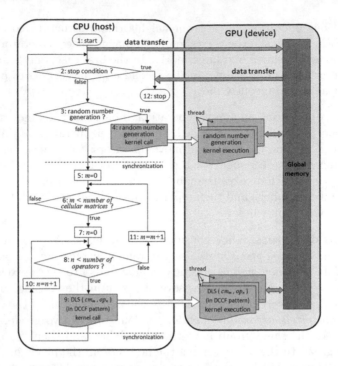

Fig. 2. Flowchart of DLS implementation.

on GPU are: the random number generation kernel and the DLS kernel. On GPU side, random numbers are needed for random move operators. The random numbers are generated in advance by the random number generation kernel which is regularly called during the algorithm according to the random number generation rate. It is the CPU side that controls DLS kernel calls with different operators executed within the DCCF pattern for frontier cells management. With several neighborhood operators in hand, we use them under the VNS framework in order to enhance the solution diversification.

4   Experimental Study

We apply the DLS algorithm to stereo matching, viewing the problem as energy minimization problem. We follow in the footsteps of Boykov et al. [14], Tappen and Freeman [18], and Szeliski et al. [15], using a simple energy function, applied to benchmark images from the widely used Middlebury stereo data set [19]. The labels are the disparities, and the data costs are the absolute color differences between corresponding pixels for each disparity. For the smoothness term in the energy function, we use a truncated linear cost as the piecewise smooth prior defined in [13]. We focus on the performance of DLS when input size augments. We experiment on the Middlebury 2005 stereo benchmark [19] including 18 pairs of images with sizes from the smallest 458×370 to the largest 1374×1110 in average. We uniformly set the disparity range to 64 pixels, for all the sizes. We denote our DLS GPU implementation as DLS-gpu. We also test the counterpart CPU sequential version which is denoted by DLS-cpu. We compare DLS with six other methods[1]: *iterated conditional modes* (ICM) [16] which is an old app-roach using a deterministic "greedy" strategy to find a local minimum; sequential *tree-reweighted message passing* (TRW-S) [15] which is an improved version of the original tree-reweighted message passing algorithm [20]; BP-S and BP-M [15] which are two updated version of the max-product *loopy belief propagation* (LBP) implementation of [18]; GC-swap and GC-expansion which are two graph cuts based algorithms proposed in [14]. Instead of reporting the absolute energy values, we report the percentage deviation from the best known solution (lowest energy) of the mean solution value over 10 runs, denoted as %*PDM* value. We choose the best known solution from the executions of all tested methods.

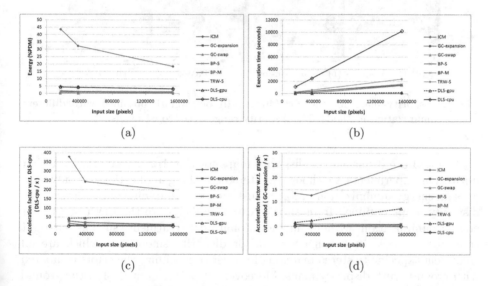

Fig. 3. Results of eight tested methods: (a) energy value as %*PDM*; (b) execution time, (c) acceleration factor of each method relative to the slowest method (DLS-cpu); (d) acceleration factor of each method relative to the method (GC-expansion) that gets the lowest energy.

The results of different methods are reported in Fig. 3. From (a) to (d) are respectively reported energy value as %*PDM*, execution time, acceleration fac-tor of each method relative to the slowest method (DLS-cpu), and acceleration factor of each method relative to the method (GC-expansion) that gets the low-est energy. The ICM method runs fastest but generates very high energies, while

---

[1] For all the tested energy minimization algorithms, we use the original codes from *http://vision.middlebury.edu/MRF/code/*.

DLS-gpu runs a little slower than ICM but generates much lower energies with more acceptable %PDM values smaller than 5%. An important observation from Fig. 3 is that, among all the tested methods, only the DLS-gpu has an acceleration factor which increases according to the augmentation of input size. This means that further improvement could be carried on only by the use of multi-processor platform with more effective cores.

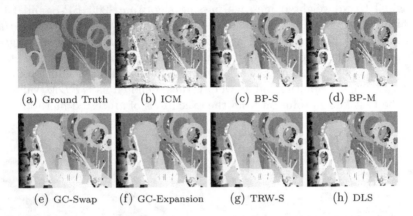

(a) Ground Truth   (b) ICM   (c) BP-S   (d) BP-M

(e) GC-Swap   (f) GC-Expansion   (g) TRW-S   (h) DLS

Fig. 4. Disparity maps for the *Art* (463×370) benchmark obtained with different energy minimization methods. The disparity range is set to 64 pixels.

In Fig. 4 are displayed the disparity maps for the *Art* benchmark. Note that during our experiments, we choose the stereo matching application but only view it as an energy minimization problem, just focusing on minimizing energies. The disparity maps obtained from all the tested methods are the raw results after energy minimization, without any additional post-treatments such as left-right consistency check, occlusion detection, or disparity smoothing, which are all treatments specific to stereo matching in order to minimize the errors compared with ground truth disparity maps. Moreover, as pointed out in [15], the ground truth solution may not always be strictly related to the lowest energy.

5   Conclusion

We have proposed a parallel formulation of local search procedure, called distributed local search (DLS) algorithm. We have applied the algorithm to stereo matching problem. The main encouraging result is that the GPU implementation of DLS on stereo matching seems to be the only method that provides an increasing acceleration factor as the instance size augments, for a result of quality less than 5% deviation to the best known energy value. For all the other approaches, the acceleration factor, against the slowest sequential version of DLS, is decreasing, except for the ICM method, which however only produces poor result of about 45% deviation to the best known energy. Graph cuts based algorithms and belief propagation based algorithms are well-performing approaches concerning quality, however the computation time increases quickly along with the instance size. That is why we hope for further improvements or improved accelerations of the DLS approach with the availability of new multi-processor platforms with more independent cores.

It is a well-known fact that the minimum energy level does not necessarily correlate to the best real-case matching. Here, we only address energy minimization discarding too much complex post-treatments necessary for the "true" ground truth matching. It should follow that many tricks are certainly not yet implemented to make energy minimization coincide to ground truth evaluation. In order to improve the matching quality in terms of minimizing the errors to ground truth only, specially designed terms for detecting typical situations in vision, such as occlusion, slanted surfaces, and the aperture problem, need to be added in the formulation of energy function.

References

1. Talbi, E.G.: Metaheuristics: From Design to Implementation, vol. 74. Wiley, Hoboken (2009)
2. Van Luong, T., Melab, N., Talbi, E.G.: Gpu computing for parallel local search metaheuristic algorithms. IEEE Trans. Comput. **62**, 173–185 (2013)
3. Delévacq, A., Delisle, P., Krajecki, M.: Parallel gpu implementation of iterated local search for the travelling salesman problem. In: Hamadi, Y., Schoenauer, M. (eds.) LION 6. LNCS, vol. 7219, pp. 372–377. Springer, Heidelberg (2012)
4. Fosin, J., Davidović, D., Carić, T.: A gpu implementation of local search operators for symmetric travelling salesman problem. PROMET Traffic Transp. **25**, 225–234 (2013)
5. Luong, T., Melab, N., Talbi, E.-G.: GPU-based multi-start local search algorithms. In: Coello, C.A.C. (ed.) LION 2011. LNCS, vol. 6683, pp. 321–335. Springer, Heidelberg (2011). doi:10.1007/978-3-642-25566-3_24
6. Sánchez-Oro, J., Sevaux, M., Rossi, A., Martí, R., Duarte, A.: Solving dynamic memory allocation problems in embedded systems with parallel variable neighborhood search strategies. Electron. Notes Discrete Math. **47**, 85–92 (2015)
7. Bengoetxea, E.: Inexact graph matching using estimation of distribution algorithms. Ph.D. thesis, Ecole Nationale Supérieure des Télécommunications, Paris, France (2002)
8. Caetano, T.S., McAuley, J.J., Cheng, L., Le, Q.V., Smola, A.J.: Learning graph matching. IEEE Trans. Pattern Anal. Mach. Intell. **31**, 1048–1058 (2009)
9. Keysers, D., Unger, W.: Elastic image matching is np-complete. Pattern Recogn. Lett. **24**, 445–453 (2003)
10. Durbin, R., Willshaw, D.: An analogue approach to the travelling salesman problem using an elastic net method. Nature **326**, 689–691 (1987)
11. Créput, J.C., Hajjam, A., Koukam, A., Kuhn, O.: Self-organizing maps in population based metaheuristic to the dynamic vehicle routing problem. J. Comb. Optim. **24**, 437–458 (2012)
12. Wang, H.: Cellular matrix for parallel k-means and local search to Euclidean grid matching. Ph.D. thesis, Université de Technologie de Belfort-Montbeliard (2015)

13. Veksler, O.: Efficient graph-based energy minimization methods in computer vision. Ph.D. thesis, Cornell University (1999)
14. Boykov, Y., Veksler, O., Zabih, R.: Fast approximate energy minimization via graph cuts. IEEE Trans. Pattern Anal. Mach. Intell. **23**, 1222–1239 (2001)
15. Szeliski, R., Zabih, R., Scharstein, D., Veksler, O., Kolmogorov, V., Agarwala, A., Tappen, M., Rother, C.: A comparative study of energy minimization methods for markov random fields with smoothness-based priors. IEEE Trans. Pattern Anal. Mach. Intell. **30**, 1068–1080 (2008)
16. Besag, J.: On the statistical analysis of dirty pictures. J. Roy. Stat. Soc. Ser. B (Methodological) **48**(3), 259–302 (1986)
17. Geman, S., Geman, D.: Stochastic relaxation, gibbs distributions, and the bayesian restoration of images. IEEE Trans. Pattern Anal. Mach. Intell. **6**, 721–741 (1984)
18. Tappen, M.F., Freeman, W.T.: Comparison of graph cuts with belief propagation for stereo, using identical mrf parameters. In: 2003 Ninth IEEE International Conference on Computer Vision. IEEE (2003)
19. Scharstein, D., Szeliski, R.: High-accuracy stereo depth maps using structured light. In: 2003 IEEE Conference on Computer Vision and Pattern Recognition, vol. 1, pp. 195–202. IEEE (2003)
20. Wainwright, M.J., Jaakkola, T.S., Willsky, A.S.: Map estimation via agreement on trees: message-passing and linear programming. IEEE Trans. Inf. Theor. **51**, 3697–3717 (2005)

Fast Hybrid BSA-DE-SA Algorithm on GPU

Mathieu Brévilliers[(✉)], Omar Abdelkafi, Julien Lepagnot,
and Lhassane Idoumghar

Université de Haute-Alsace (UHA), LMIA (E.A. 3993),
4 rue des frères Lumière, 68093 Mulhouse, France
{mathieu.brevilliers,omar.abdelkafi,julien.lepagnot,
lhassane.idoumghar}@uha.fr

Abstract. This paper introduces a hybridization of Backtracking
Search Optimization Algorithm (BSA) with Differential Evolution (DE)
and Simulated Annealing (SA) in order to improve the convergence speed
of BSA. An experimental study, conducted on 20 benchmark problems,
shows that this approach outperforms BSA and two other hybridizations [4,18], in terms of solution quality and convergence speed. We
also describe our CUDA implementation of this algorithm for graphics processing unit (GPU). Experimental results are reported for 10
high-dimensional benchmark problems, and it highlights that significant
speedup can be achieved.

Keywords: Continuous optimization · Hybrid metaheuristic · Backtracking search optimization algorithm · Differential evolution ·
Simulated annealing · Graphics processing unit · CUDA

1 Introduction

Optimization consists in finding the set of parameters that leads to the best possible value for a given cost function. Metaheuristics refer to a class of methods
that are often inspired by analogies (for example, with physics or biology), and
that are designed to solve hard optimization problems without any knowledge of
the practical context. In this class, evolutionary algorithms are population-based
algorithms that use evolution mechanisms (such as mutation and crossover),
in order to approximate the best solution. Following this idea, several efficient
approaches have emerged, such as artificial bee colony algorithms [8], particle
swarm optimization algorithms [6,19], the covariance matrix adaptation evolution strategy [5], or differential evolution algorithms [15,17].

Among all existing evolutionary strategies, it has been shown that Backtracking Search Optimization Algorithm (BSA) [3] can also find high-quality solutions
for a large number of continuous optimization problems. Several extensions of
this algorithm have been proposed to improve either solution quality or convergence speed [2,18]. As BSA mainly focuses on exploration, it can be quite slow
converging on the global best solution, and it would be challenging to speed up
its convergence without loss of quality.

© Springer International Publishing AG 2016
P. Siarry et al. (Eds.): ICSIBO 2016, LNCS 10103, pp. 75–86, 2016.
DOI: 10.1007/978-3-319-50307-3_6

In this context, we present a hybrid algorithm that uses differential evolution (DE) and simulated annealing (SA) techniques together with BSA principles. The aim of this work is to speed up the convergence of BSA, that is to significantly reduce the number of function evaluations needed to achieve high-quality solutions. We also propose an implementation for graphics processing unit (GPU) to investigate the benefit in terms of runtime speedup for high-dimensional instances.

Section 2 presents the BSA algorithm and two BSA-DE hybridizations proposed in the literature. Section 3 introduces our BSA-DE-SA hybrid approach and experimental results are reported to show the superiority of this approach. The design of our algorithm for a GPU implementation is described in Sect. 4, and an experimental study shows to what extent the execution of the algorithm can be accelerated. Finally, concluding remarks and perspectives are given in Sect. 5.

2   Related Work

2.1   Backtracking Search Optimization Algorithm

Backtracking Search Optimization Algorithm (BSA) [3] is an evolutionary algorithm for continuous optimization. It has a classical structure where a population evolves from generation to generation according to the operators listed below. In the following, let N denote the number of individuals in the population, and D the number of dimensions in the considered optimization problem.

1. Selection-I. As a backtracking strategy, BSA has a memory to store a historical population, that consists of the individuals of a previous generation. Selection-I updates this memory with probability 0.5, by replacing the whole historical population with a random permutation of the current population.

2. Mutation. A new mutant population M is created from the current population P and from the historical population $oldP$ by using the following equation:

$$\forall i \in \{1, ..., N\}, \forall j \in \{1, ..., D\}, M_{i,j} = P_{i,j} + F^{BSA} \times (oldP_{i,j} - P_{i,j}) \qquad (1)$$

where $F^{BSA} = 3 \times randn$, and $randn$ is a real value randomly generated with the standard normal distribution. A new value of F^{BSA} is generated for each generation.

3. Crossover. Two crossover strategies are randomly used (with probability 0.5) to get a new trial population T from M and P. The first strategy depends on a user-defined parameter called *mixrate*, that controls how many dimensions (at most) of a mutant individual will be incorporated in a trial individual. The second strategy ensures that, for each trial individual, only one randomly chosen dimension will come from the corresponding mutant individual.

4. Confinement. The boundary control mechanism checks if any trial individual is outside the search space (due to Eq. 1). In such a case, the concerned dimensions are randomly regenerated inside the appropriate bounds.

5. Selection-II. Each trial individual T_i is evaluated and, if T_i is better than P_i, then P_i is replaced with T_i in P.

Experimental study and statistical tests [3] have shown that BSA is generally better than SPSO2011 [19], CMAES [5], ABC [8], JDE [1], CLPSO [9], and SADE [14]. The main advantages of BSA are that it has few user-defined parameters (the population size N, and *mixrate*) and that it can solve a wide range of optimization problems, due to its good exploration ability. However, BSA main shortcoming is that it can be quite slow converging on the global best solution. Since then, BSA has been applied, improved, or hybridized in several ways [2,4,10,16,18].

2.2 Hybrid BSA-DE Algorithms

We present here two hybridizations that inspired the algorithm proposed in Sect. 3, and that will be used for comparison.

BSA-DE. Das et al. [4] replaced Eq. 1 of BSA in the following way, by adding a term coming from the DE/target-to-best/1 mutation scheme [13]:

$$\forall i \in \{1, ..., N\}, \forall j \in \{1, ..., D\},$$
$$M_{i,j} = P_{i,j} + F^{BSA} \times (oldP_{i,j} - P_{i,j}) + F^{DE} \times (P_{best,j} - P_{i,j}) \quad (2)$$

where F^{BSA} is defined as in Eq. 1, F^{DE} is the scaling factor of DE, and *best* $\in \{1, ..., N\}$ is the index of the best individual in P. In contrast with BSA, a new value of F^{BSA} is generated for each individual. It has been shown that this BSA-DE hybridization generally performs better and converges faster than BSA and DE.

HBD. Wang et al. [18] proposed a hybridization where DE follows BSA in the generation loop: for each generation, a BSA iteration is firstly applied, and then DE is applied to improve only 1 bad individual of the current population. This bad individual is randomly chosen with respect to its fitness: the worse the fitness, the higher the probability. Then, the DE/best/1 mutation scheme [15] and a binomial crossover are used to generate a trial individual, that will replace the current individual if it performs better. Comparing this so-called HBD algorithm with BSA, it has been shown that HBD outperforms BSA in terms of solution quality and convergence speed.

3 Contribution to Speed up BSA Convergence

3.1 Hybrid BSA-DE-SA Algorithm

The proposed hybrid approach is based on a two-level BSA-DE combination and on a SA schedule to gradually decrease the range of BSA scaling factor. The aim is to improve the convergence of the basic BSA algorithm. More precisely, we merge the two ideas given in Sect. 2.2: for each generation, a BSA iteration with a DE-inspired mutation equation is firstly applied (individual-level of hybridization), and then DE is used to optimize a few bad individuals in the population (generation-level of hybridization).

Individual-level BSA-DE hybridization. We define 2 new scaling factors. The first one, called *intensification factor*, and denoted F^I, is defined by the user in $[0, 1]$. The second one, called *exploration factor*, and denoted F_i^E, is generated for each individual i during the mutation process:

$$\forall i \in \{1, ..., N\}, \; F_i^E = C \times randn, \tag{3}$$

where C is a coefficient decreasing with time from generation to generation (see below), and *randn* is a real value randomly generated with the standard normal distribution. Then, Eq. 1 is modified as follows, in a slightly different way from [4], in order to instill the DE/target-to-best/1 scheme into BSA mutation operator:

$$\forall i \in \{1, ..., N\}, \forall j \in \{1, ..., D\},$$
$$M_{i,j} = P_{i,j} + F_i \times (oldP_{i,j} - oldP_{k,j}) + F^{DE} \times (P_{best,j} - P_{i,j}), \tag{4}$$

where k is randomly chosen in $\{1, ..., N\}$ such that $k \neq i$. The factor F_i replaces F^{BSA}, and is defined by the equation:

$$F_i = \begin{cases} F_i^E & \text{if } rand > \frac{1}{16}, \\ F^I & \text{otherwise,} \end{cases} \tag{5}$$

where *rand* is a random value uniformly generated in $[0, 1]$.

SA schedule for C. According to the temperature cooling schedule in SA, the coefficient C is gradually decreased from 3 to 1 with a geometric law during the first third of the algorithm (in terms of number of function evaluations).

Generation-level BSA-DE hybridization. The HBD method proposed in [18] is applied after each iteration of the individual-level BSA-DE hybridization. For reasons of scalability, the number of bad individuals selected to be optimized in this stage is related to the population size: 1 individual is selected if $N < 30$, and $\lfloor N/30 \rfloor$ otherwise.

Equation 5 together with the range of C and F^I show that a few individuals are used to intensify the search with a low F_i, while the major part explores the search space with a larger F_i. Furthermore, the SA schedule for decreasing C allows to use the full exploration ability of the algorithm at the beginning, and to develop its exploitation ability at a later stage. Finally, the two-level BSA-DE hybridization allows to combine in the same algorithm the DE/best/1 scheme (generation-level) with a DE/target-to-best/1-like scheme (individual-level), in order to speed up the convergence of the algorithm.

3.2   Experimental Results

We realized an experimental study in order to compare our hybrid BSA-DE-SA approach with BSA [3], BSA-DE [4], and HBD [18]. Specifically, two versions of BSA-DE-SA have been implemented: BDS-1 that only uses the individual-level BSA-DE hybridization with a SA schedule for C, and BDS-2 that uses all fea-

Table 1. List of benchmark problems (ID: function identifier; Low, Up: limits of search space; D: dimension).

ID	Name	Low	Up	D
F1	Schwefel 1.2	−100	100	30
F2	Schwefel 2.22	−10	10	30
F3	Sphere	−100	100	30
F4	Ackley	−32	32	30
F5	Griewank	−600	600	30
F6	Rastrigin	−5.12	5.12	30
F7	Rosenbrock	−30	30	30
F8	Schaffer f6	−100	100	2
F9	Weierstrass	−0.5	0.5	10
F10	Shifted sphere	−100	100	10
F11	Shifted Schwefel 1.2	−100	100	10
F12	Shifted rotated high conditioned elliptic function	−100	100	10
F13	Shifted Schwefel 1.2 with noise	−100	100	10
F14	Schwefel 2.6	−100	100	10
F15	Shifted Rosenbrock	−100	100	10
F16	Shifted rotated Griewank	0	600	10
F17	Shifted rotated Ackley	−32	32	10
F18	Shifted Rastrigin	−5	5	10
F19	Shifted rotated Rastrigin	−5	5	10
F20	Shifted rotated Weierstrass	−0.5	0.5	10

Table 2. Control parameter settings for the compared algorithms.

Algorithm	Parameters
BSA [3]	$N = 30$, $mixrate = 1$
BSA-DE [4]	$N = 30$, $mixrate = 1$, $F^{DE} = 0.5$
HBD [18]	$N = 30$, $mixrate = 1$, scaling factor $F = 0.8$, crossover rate $C_r = 0.9$, DE applied on $\lfloor N/30 \rfloor = 1$ individual
BDS-1	$N = 30$, $mixrate = 1$, $F^{DE} = 0.5$, $F^I = 0.5$ applied for each individual with probability $1/16$, C decreased from 3 to 1 during the first $1/3$ of the allowed function evaluations
BDS-2	BDS-1 settings together with HBD settings

Table 3. Basic statistics of the two versions of BSA-DE-SA, and comparison with BSA [3], BSA-DE [4], and HBD [18] (Mean: mean error; Std: standard deviation; Best: best error). Best values are depicted in bold font.

ID	Statistics	BDS-1	BDS-2	BSA [3]	BSA-DE [4]	HBD [18]
F1	Mean	0	0	3.45331725e-1	0	4.69223633e-5
	Std	0	0	3.56207055e-1	0	4.87788549e-5
	Best	0	0	4.65828600e-2	0	1.74837295e-6
F2	Mean	0	0	0	0	0
	Std	0	0	0	0	0
	Best	0	0	0	0	0
F3	Mean	0	0	0	0	0
	Std	0	0	0	0	0
	Best	0	0	0	0	0
F4	Mean	0	0	0	0	0
	Std	0	0	0	0	0
	Best	0	0	0	0	0
F5	Mean	2.21758219e-3	1.97145704e-3	0	7.55217402e-3	4.93069355e-4
	Std	4.81315491e-3	4.16365025e-3	0	7.53723631e-3	1.87643557e-3
	Best	0	0	0	0	0
F6	Mean	0	0	3.31653019e-2	0	1.65826509e-1
	Std	0	0	1.81653839e-1	0	5.27993560e-1
	Best	0	0	0	0	0
F7	Mean	9.30325416e-1	1.32887461	2.35616889e+1	**6.64437376e-01**	8.01149354e-1
	Std	1.71491464	1.91143983	2.90306080e+1	**1.51112585**	1.62101635
	Best	0	0	5.31405876e-7	0	0
F8	Mean	3.37430650e-4	**1.09433786e-6**	2.73248252e-3	7.65013736e-4	2.60244739e-3
	Std	1.77204503e-3	**5.99392266e-6**	4.04657848e-3	2.47492459e-3	4.36309515e-3
	Best	0	0	0	0	0
F9	Mean	0	0	0	0	0
	Std	0	0	0	0	0
	Best	0	0	0	0	0
F10	Mean	0	0	0	0	0
	Std	0	0	0	0	0
	Best	0	0	0	0	0
F11	Mean	0	0	8.12184166e-7	0	0
	Std	0	0	1.18619825e-6	0	0
	Best	0	0	0	0	0
F12	Mean	1.88063034e+3	**6.70067111e+2**	1.62772681e+4	6.63797991e+3	5.12822952e+3
	Std	4.09511408e+3	**8.99497851e+2**	2.63103587e+4	5.96963034e+3	6.89120964e+3
	Best	6.85806410	**1.69290665e-1**	3.23132561e+2	1.23221979e+2	1.28697388e+1
F13	Mean	0	0	3.52038638e-3	0	0
	Std	0	0	1.00832481e-2	0	1.41395434e-8
	Best	0	0	1.16021564e-5	0	0
F14	Mean	0	0	1.63586845e-2	0	5.28382701e-5
	Std	0	0	3.29592107e-2	0	6.56037241e-5
	Best	0	0	1.06714993e-4	0	2.68750955e-6
F15	Mean	1.32885971e-1	0	2.31962945e-1	1.32889360e-1	5.79353282e-4
	Std	7.27846435e-1	0	5.86248030e-1	7.27845795e-1	3.01367607e-3
	Best	0	0	0	0	0
F16	Mean	5.42895964e-2	4.61309502e-2	6.56037488e-2	1.14081123e-1	**3.33610373e-2**
	Std	4.71316146e-2	2.29246572e-2	3.49897515e-2	5.14108950e-2	**2.15975637e-2**
	Best	7.52199899e-3	9.85728587e-3	0	0	0
F17	Mean	2.03415389e+1	2.03230528e+1	**2.03225585e+1**	2.03462701e+1	2.03325172e+1
	Std	**7.02011419e-2**	8.34903645e-2	8.21386118e-2	7.14983620e-2	7.80534782e-2
	Best	2.01888263e+1	**2.00865221e+1**	2.01202686e+1	2.02124186e+1	2.02032472e+1
F18	Mean	0	0	0	0	0
	Std	0	0	0	0	0
	Best	0	0	0	0	0
F19	Mean	6.84981566	6.49766160	1.14543047e+1	**5.72620112**	1.10771999e+1
	Std	2.71890009	2.90703990	4.08944890	**2.40672118**	3.94305927
	Best	2.25212064	**1.98991811**	4.97479545	2.98487717	4.97479528
F20	Mean	4.83483045	4.47945956	4.39522547	4.36457606	**2.89661069**
	Std	**9.79152736e-1**	1.30055492	1.11792825	1.49610439	1.31134257
	Best	2.66166689	1.15061555	1.32751014	4.46557673e-1	**3.08726178e-1**

tures described above. All these algorithms have been tested on the benchmark functions listed in Table 1, and Table 2 shows the values of the control parameters for each algorithm. Each algorithm has been run 30 times on each benchmark function. $10\,000 \times D$ function evaluations per run are allowed, and a benchmark problem is considered as solved when a fitness lower than $f_{opt} + 10^{-8}$ is reached, where f_{opt} denotes the corresponding optimal fitness.

Table 3 reports basic statistics for the compared algorithms. We can see that BDS-2 gets 14 times the first place in terms of mean error, whereas BSA-DE, BDS-1, HBD and BSA make it respectively 13, 11, 10, and 8 times. BDS-2

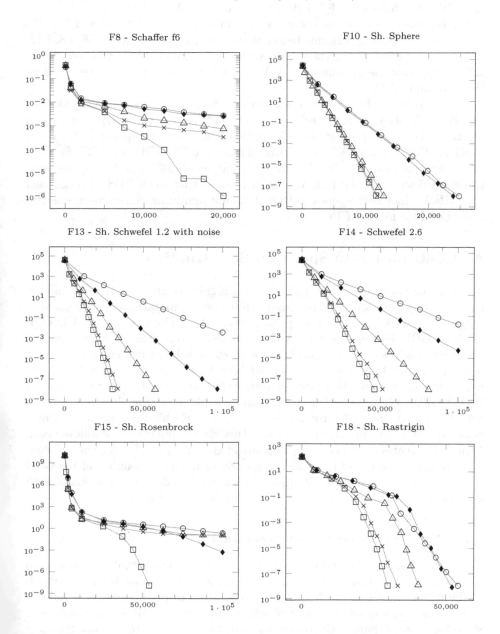

Fig. 1. The curves show how many function evaluations (x-axis) are needed to reach a certain mean error (y-axis in log scale) for selected benchmark problems of Table 1. BSA is depicted with empty circles, BSA-DE with empty triangles, HBD with filled diamonds, BDS-1 with crosses, and BDS-2 with empty squares.

beats BSA on 11 functions (F1, F6–8, F11–16, F19), HBD on 8 functions (F1, F6, F8, F12, F14, F15, F17, F19), and BSA-DE on 6 functions (F5, F8, F12, F15–17). Conversely, BDS-2 loses to HBD on 4 functions (F5, F7, F16, F20), to BSA-DE on 3 functions (F7, F19, F20), and to BSA on 3 functions (F5, F17, F20). We can notice similar results when comparing BDS-1 to BSA, BSA-DE, and HBD, but BDS-2 performs better than BDS-1 on 8 functions (F5, F8, F12, F15–17, F19, F20). From these observations, we can conclude that our BSA-DE-SA approach clearly outperforms BSA, and gives slightly better results than BSA-DE and HBD.

Figure 1 shows the convergence curves for some of the benchmark problems and it highlights that our hybrid approach leads to faster convergence: we can see that BDS-2 saves about 50% of function evaluations compared to BSA and HBD for F10, between 45% and 70% compared to BSA-DE and HBD for F13, about 40% compared to BSA-DE for F14, and between 25% and 45% compared to BSA-DE and HBD for F18. We can also notice that when BDS-1 is combined with HBD, which corresponds to BDS-2, then a significant convergence speedup is obtained for F8 and F15.

4   Contribution to Speed up BSA Runtime

The graphics processing unit (GPU) has a highly parallel architecture, and it can be easily programmed for general purpose computations with high-level languages, thanks to dedicated parallel computing platforms like CUDA for NVIDIA GPU devices. The CUDA platform allows to realize heterogeneous parallel computations, which means that the program is launched on the CPU, that delegates parallel subroutines (so-called kernels) to the GPU. In CUDA programming, each kernel is a piece of code called from the CPU and duplicated on the GPU to be executed in parallel on multiple data (the GPU has a SIMD architecture, i.e. single-instruction multiple-data). Each kernel duplicate is executed by a CUDA thread, and all these threads are organized as follows: each kernel call creates a grid composed of thread groups, called blocks, that all contain the same number of threads. Thus, in order to take advantage of the GPU performance, any evolutionary algorithm should be adapted, in terms of data decomposition, to be processed in parallel by blocks of threads [7,11,12].

4.1   Design of the GPU Implementation

The first feature of our proposed CUDA implementation is that we delegate to the GPU the most time-consuming part of the algorithm, that is the evaluation of the population. This can be done with two levels of parallelization as follows. Firstly, the evaluations of all individuals can be done in parallel. And secondly, since for the most part of the benchmark functions we need to perform the same computations on each dimension before aggregating the results (for example, with a sum), the dimensions can also be processed in parallel. Getting back to

CUDA programming, it means that the evaluation workload can be divided into N blocks of D threads, that each deals with 1 dimension of 1 individual.

However, as already noticed in the literature [12], if the evaluation is the only task entrusted to the GPU, the algorithm has to transfer the whole population from CPU memory to GPU in every generation, which is very slow compared to arithmetic computations on GPU. Therefore, we choose to store the population in the GPU global memory in order to minimize the time lost in data transfer. It means that all steps of the algorithm are processed by the GPU, while the generation loop is done by the CPU, that launches a GPU kernel for each step with the ad-hoc data decomposition (in terms of CUDA blocks and threads). As much as possible, we divide the processings into N blocks of D threads: as seen above, this is particularly suited to evaluate the population, but also, for example, to generate the initial population, to apply the mutation equation, or to perform the boundary control. In addition to that, other decompositions are sometimes needed, depending on the processing to be realized: for example, 1 block of N threads to find the best individual, or 1 block of D threads to update the global best solution.

4.2   Experimental Results

We realized an experimental study in order to compare our GPU implementations of BDS-1 and BDS-2 with their sequential versions and with the original sequential BSA [3]. For reasons of dimensional scalability, these algorithms have been tested on the benchmark functions F1–9 of Table 1 and on Michalewics function (denoted as F21, and defined on $[0, 3.1416]$, according to [3]). The control parameters of each algorithm have been set as shown in Table 2, except the population size that now depends on the problem dimension as follows: $N = D$. Several experiments have been conducted with $D = 128$, $D = 256$, and $D = 512$. For a given value of D, each algorithm has been run 15 times on each benchmark problem, and $3\,000 \times D$ function evaluations per run were allowed. For these experimentations, all the compared algorithms are written in C/C++, and the corresponding programs are compiled on an Intel Core processor i5-3330 CPU (3.00 GHz) with 4 GB of RAM and a NVIDIA GeForce GTX680 GPU.

Table 4 reports basic statistics for the compared algorithms. First of all, it seems that BSA tends to find better solutions than BDS-1 and BDS-2 when N and D increase: BSA beats BSA-DE-SA approaches on 3 functions out of 10 when $N = D = 128$, on 4 functions when $N = D = 256$, and on 6 functions when $N = D = 512$. However, all compared results almost always have the same order of magnitude. Secondly, we can see that BDS-2 clearly outperforms BDS-1 in terms of solution quality: both CPU and GPU versions of BDS-2 win against BDS-1 on 20 experiments out of the 30 listed in Table 4. Thirdly, the resulting mean runtimes show that BDS-1 GPU version can lead up to a 40 time speedup with regard to BDS-1 CPU version. It sounds that the acceleration mainly comes from the evaluation of the population, and that it directly depends on the computation complexity of the considered benchmark function. Fourthly, we can notice that BDS-2 speedup is much lower than that of BDS-1. It is due to the

Table 4. Comparison of BSA [3], BDS-1, and BDS-2 in high dimensions (Mean: mean solution; Time: mean runtime in seconds). Best values are depicted in bold font.

N=D	ID	Statistics	BSA [3] CPU	BDS-1 CPU	BDS-1 GPU	BDS-1 Speedup	BDS-2 CPU	BDS-2 GPU	BDS-2 Speedup
128	F1	Mean	3.2531e+3	2.3844e+3	2.6854e+3		**1.9063e+3**	2.0242e+3	
		Time	11.13	11.25	**2.89**	3.90	11.53	19.49	0.59
	F2	Mean	1.3430e-1	2.1966e-3	9.4546e-3		3.4180e-3	**1.5114e-3**	
		Time	**1.94**	2.17	2.94	0.74	2.33	19.43	0.12
	F3	Mean	2.2080e-1	3.6920e-8	4.3224e-8		3.4427e-10	**3.0185e-10**	
		Time	**1.89**	2.08	2.89	0.72	2.24	19.35	0.12
	F4	Mean	**4.5019e-2**	2.6885	2.7312		2.4161	2.5679	
		Time	3.41	3.61	**2.97**	1.22	3.71	19.45	0.19
	F5	Mean	8.2368e-2	8.6823e-3	3.9421e-3		**3.7734e-3**	3.7751e-3	
		Time	3.82	3.80	**2.98**	1.27	3.93	19.46	0.20
	F6	Mean	1.6949e+2	1.1462e+2	**1.1045e+2**		1.2092e+2	1.2230e+2	
		Time	3.80	3.88	**2.64**	1.47	4.07	19.58	0.21
	F7	Mean	5.2074e+2	3.5942e+2	3.2152e+2		2.7981e+2	**2.5426e+2**	
		Time	**2.87**	3.05	3.03	1.01	3.23	19.50	0.17
	F8	Mean	4.5822e-1	4.6930e-1	4.6597e-1		4.6448e-1	4.6425e-1	
		Time	**1.90**	2.04	2.64	0.77	2.21	19.24	0.11
	F9	Mean	**1.2849**	1.1354e+1	1.1646e+1		1.1436e+1	1.1518e+1	
		Time	63.67	64.41	**3.71**	17.37	64.80	20.39	3.18
	F21	Mean	-9.3241e+1	**-9.8617e+1**	-9.8456e+1		-9.8065e+1	-9.8098e+1	
		Time	9.34	9.45	**2.74**	3.45	9.65	19.65	0.49
256	F1	Mean	**7.4732e+4**	1.5148e+4	1.5590e+4		1.4160e+4	1.4756e+4	
		Time	80.87	81.20	**6.49**	12.51	82.47	73.08	1.13
	F2	Mean	3.4245	1.5047	1.6753		**8.2807e-1**	1.2669	
		Time	7.73	8.43	**6.16**	1.37	9.04	72.78	0.12
	F3	Mean	3.5056e+1	1.9425e-1	1.1156e-1		1.2587e-2	**4.2498e-3**	
		Time	7.39	8.11	**6.09**	1.33	8.74	72.68	0.12
	F4	Mean	**1.1330**	4.4814	4.6626		4.4919	4.2836	
		Time	14.96	15.04	**6.35**	2.37	15.65	72.97	0.21
	F5	Mean	1.3237	4.5771e-2	5.4828e-2		2.1115e-2	**1.5127e-2**	
		Time	15.56	15.35	**6.39**	2.40	15.75	73.00	0.22
	F6	Mean	6.2993e+2	6.0159e+2	**5.9217e+2**		6.1518e+2	6.2133e+2	
		Time	15.23	15.63	**5.63**	2.78	16.33	73.55	0.22
	F7	Mean	2.0790e+3	1.2885e+3	1.3246e+3		**8.8510e+2**	9.3211e+2	
		Time	11.38	12.02	**6.58**	1.83	12.70	73.09	0.17
	F8	Mean	**4.8217e-1**	4.9603e-1	4.9605e-1		4.9582e-1	4.9573e-1	
		Time	7.36	7.90	**5.48**	1.44	8.51	72.23	0.12
	F9	Mean	**1.3822e+1**	7.6411e+1	7.6420e+1		7.7453e+1	7.7002e+1	
		Time	256.59	262.90	**8.65**	30.40	263.94	75.39	3.50
	F21	Mean	-1.4556e+2	**-1.5584e+2**	-1.5502e+2		-1.5358e+2	-1.5336e+2	
		Time	37.74	37.57	**5.97**	6.29	38.42	73.97	0.52
512	F1	Mean	**1.2895e+4**	6.0397e+4	6.6561e+4		5.8543e+4	5.9229e+4	
		Time	613.77	614.39	**20.46**	30.02	620.01	289.71	2.14
	F2	Mean	**2.4369e+1**	3.8593e+1	4.5054e+1		3.0485e+1	4.0408e+1	
		Time	31.32	33.58	**17.45**	1.92	36.01	286.69	0.13
	F3	Mean	5.0457e+2	1.6084e+2	1.3988e+2		9.6491e+1	**5.4097e+1**	
		Time	29.80	32.26	**16.41**	1.97	34.78	285.44	0.12
	F4	Mean	**2.7341**	7.2895	7.0967		6.9065	6.9606	
		Time	61.69	61.76	**18.05**	3.42	63.96	286.98	0.22
	F5	Mean	5.1719	2.4347	2.3131		**1.5618**	1.6490	
		Time	61.89	63.38	**18.21**	3.48	64.72	287.26	0.23
	F6	Mean	**1.8488e+3**	2.0890e+3	1.9805e+3		2.1301e+3	1.8501e+3	
		Time	60.65	63.28	**15.60**	4.06	66.01	286.14	0.23
	F7	Mean	8.4335e+3	1.6159e+4	1.3098e+4		**5.9431e+3**	6.0513e+3	
		Time	45.65	47.92	**18.31**	2.62	50.77	287.76	0.18
	F8	Mean	**4.9049e-1**	4.9967e-1	4.9964e-1		4.9962e-1	4.9959e-1	
		Time	29.44	31.36	**15.43**	2.03	33.96	284.72	0.12
	F9	Mean	6.1091e+1	2.6234e+2	2.6643e+2		2.6058e+2	2.5402e+2	
		Time	1027.70	1062.34	**26.48**	40.12	1066.46	295.91	3.60
	F21	Mean	-2.1917e+2	-2.3382e+2	**-2.3394e+2**		-2.3102e+2	-2.3067e+2	
		Time	151.40	151.35	**16.87**	8.97	155.07	290.21	0.53

HBD part of BDS-2: one level of parallelization is lost in this part of the GPU algorithm, since Sect. 3.1 and Table 2 point out that all HBD evolutionary operators are applied only for a few individuals ($N/30$). So, almost all the speedup gained from BSA iteration is then lost in the DE iteration needed for the HBD part of BDS-2. In a word, we can conclude that BDS-1 GPU version seems to be the most suitable in terms of runtime speedup for the selected high-dimensional benchmark problems.

5   Conclusion

A hybrid BSA-DE-SA algorithm has been presented and an experimental study on 20 benchmark problems shows that it performs well in terms of solution quality and convergence speed. Then, the design of our GPU implementation has been explained, and experimental results point out that a significant speedup can be achieved, up to 40 times with regard to sequential program.

In future work, we will consider comparing our approach to other algorithms (for example, PSO, CMAES, SHADE) with additional benchmark functions. As we introduce new user-defined parameters, another perspective would be to improve the proposed algorithm with a self-adaptive technique, in order to be less user-dependent and to achieve possibly better results. Finally, in the longer term, it would be interesting to compare this hybridization with existing large-scale optimization methods.

References

1. Brest, J., Greiner, S., Boskovic, B., Mernik, M., Zumer, V.: Self-adapting control parameters in differential evolution: a comparative study on numerical benchmark problems. IEEE Trans. Evol. Comput. **10**(6), 646–657 (2006)
2. Brévilliers, M., Abdelkafi, O., Lepagnot, J., Idoumghar, L.: Idol-guided backtracking search optimization algorithm. In: 12th International Conference on Artificial Evolution (EA 2015), Lyon, France, October 2015
3. Civicioglu, P.: Backtracking search optimization algorithm for numerical optimization problems. Appl. Math. Comput. **219**(15), 8121–8144 (2013)
4. Das, S., Mandal, D., Kar, R., Ghoshal, S.P.: A new hybridized backtracking search optimization algorithm with differential evolution for sidelobe suppression of uniformly excited concentric circular antenna arrays. Int. J. RF Microwave Comput. Aided Eng. **25**(3), 262–268 (2015)
5. Hansen, N., Ostermeier, A.: Completely derandomized self-adaptation in evolution strategies. Evol. Comput. **9**(2), 159–195 (2001)
6. Idoumghar, L., Idrissi-Aouad, M., Melkemi, M., Schott, R.: Metropolis particle swarm optimization algorithm with mutation operator for global optimization problems. In: 2010 22nd IEEE International Conference on Tools with Artificial Intelligence (ICTAI), vol. 1, pp. 35–42, October 2010
7. Kalivarapu, V., Winer, E.: A study of graphics hardware accelerated particle swarm optimization with digital pheromones. Struct. Multidisc. Optim. **51**(6), 1281–1304 (2015)
8. Karaboga, D., Basturk, B.: A powerful and efficient algorithm for numerical function optimization: artificial bee colony (ABC) algorithm. J. Global Optim. **39**(3), 459–471 (2007)
9. Liang, J.J., Qin, A.K., Suganthan, P.N., Baskar, S.: Comprehensive learning particle swarm optimizer for global optimization of multimodal functions. IEEE Trans. Evol. Comput. **10**(3), 281–295 (2006)
10. Lin, Q., Gao, L., Li, X., Zhang, C.: A hybrid backtracking search algorithm for permutation flow-shop scheduling problem. Comput. Ind. Eng. **85**, 437–446 (2015)
11. Luo, G.-H., Huang, S.-K., Chang, Y.-S., Yuan, S.-M.: A parallel bees algorithm implementation on GPU. J. Syst. Archit. **60**(3), 271–279 (2014)

12. Pospichal, P., Jaros, J., Schwarz, J.: Parallel genetic algorithm on the CUDA architecture. In: Chio, C., et al. (eds.) EvoApplications 2010. LNCS, vol. 6024, pp. 442–451. Springer, Heidelberg (2010). doi:10.1007/978-3-642-12239-2_46
13. Price, K.: An introduction to differential evolution. In: Corne, D., Dorigo, M., Glover, F. (eds.) New Ideas in Optimization, pp. 79–108. McGraw-Hill Ltd., London (1999)
14. Qin, A.K., Suganthan, P.N.: Self-adaptive differential evolution algorithm for numerical optimization. In: The 2005 IEEE Congress on Evolutionary Computation, vol. 2, pp. 1785–1791 (2005)
15. Storn, R., Price, K.: Differential evolution: a simple and efficient heuristic for global optimization over continuous spaces. J. Global Optim. **11**(4), 341–359 (1997)
16. Syed, M.S., Injeti, S.K.: Simultaneous optimal placement of DGs and fixed capacitor banks in radial distribution systems using BSA optimization. Int. J. Comput. Appl. **108**(5), 28–35 (2014)
17. Tanabe, R., Fukunaga, A.: Success-history based parameter adaptation for differential evolution. In: 2013 IEEE Congress on Evolutionary Computation (CEC), pp. 71–78, June 2013
18. Wang, L., Zhong, Y., Yin, Y., Zhao, W., Wang, B., Xu, Y.: A hybrid backtracking search optimization algorithm with differential evolution. Math. Probl. Eng. **2015**, 16 (2015). doi:10.1155/2015/769245. Article ID 769245
19. Zambrano-Bigiarini, M., Clerc, M., Rojas, R.: Standard particle swarm optimisation: a baseline for future PSO improvements. In: 2013 IEEE Congress on Evolutionary Computation, pp. 2337–2344, June 2013

A New Parallel Memetic Algorithm to Knowledge Discovery in Data Mining

Dahmri Oualid[1][(⊠)] and Ahmed Riadh Baba-Ali[2]

[1] Computer Science Department, FEI, USTHB,
BP 32 El Alia, Bab Ezzouar, Algeria
dahmri_oualid_39@yahoo.fr
[2] Research Laboratory LRPE, FEI, USTHB,
BP 32 El Alia, Bab Ezzouar, Algeria
riadhbabaali@yahoo.fr

Abstract. This paper presents a new parallel memetic algorithm (PMA) for solving the problem of classification in the process of Data Mining. We focus our interest on accelerating the PMA. In most parallel algorithms, the tasks performed by different processors need access to shared data, this creates a need for communication, which in turn slows the performance of the PMA. In this work, we will present the design of our PMA, In which we will use a new replacement approach, which is a hybrid approach that uses both Lamarckian and Baldwinian approaches at the same time, to reduce the quantity of informations exchanged between processors and consequently to improve the speedup of the PMA. An extensive experimental study performed on the UCI Benchmarks proves the efficiency of our PMA. Also, we present the speedup analysis of the PMA.

Keywords: Parallel memetic algorithm · Classification · Extraction of rules · Lamarckian approach · Baldwinian approach · Hybridization

1 Introduction

Nowadays there is a huge amount of data being collected and stored in databases everywhere across the globe, and there are invaluable informations and knowledge "hidden" in such databases, and without automatic methods for extracting this informations, it is practically impossible to use them.

Data mining [1], was born for this need. Among the tasks of this process, we find the supervised classification [2] is one of the most important. It consists of predicting a certain outcome based on a given input. In order to predict the outcome, the algorithm processes a training set containing a set of attributes and the respective outcome, usually called goal or prediction attribute. The algorithm tries to discover relationships between the attributes that would make it possible to predict the outcome. Next, the algorithm is given a data set not seen before, called prediction set, which contains the same set of attributes, except for the prediction attribute – not yet known. The algorithm analyses the input and produces a prediction. The prediction accuracy defines how "good" the

© Springer International Publishing AG 2016
P. Siarry et al. (Eds.): ICSIBO 2016, LNCS 10103, pp. 87–101, 2016.
DOI: 10.1007/978-3-319-50307-3_7

algorithm is. This problem is NP-hard [3] and for that reason an exponential complexity making impossible the use of exact methods when the data size is large.

Meta-heuristics [4, 5] are algorithms that can provide a satisfactory solution in a relatively short time on this class of problems. Among these methods, we are particularly interested in the Memetic Algorithms [18] (hybridization of a local search [7] and genetic algorithm [6]). The genetic algorithm is so widely used to solve data mining classification problems is the fact that prediction rules are very naturally represented in GA. Additionally, GA has proven to produce good results with global search problems like classification. But this kind of algorithms requires considerable computation time and amount of memory which are closely related to the size of the problem and to the quality of the solution to obtain.

Therefore, these algorithms become interesting to parallelize. In general, parallelism is used to solve complex problems requiring expensive algorithms in terms of execution time. But in most parallel algorithms, the tasks performed by different processors need access to shared data, this creates a need for communication which in turn slows the performance of the parallel algorithm. These communications are even more influential, in the case where processors require data generated by other processors. So the objective of this work is to minimize communications in terms of data volume and frequency of exchanges without penalizing the quality of the solution.

2   Related Work

Genetic Algorithms are those among which have been the subject of the greatest number of parallelization work, particularly because of their fundamental parallel nature [8]. Cantú-Paz [9] presented a review of the main publications related to parallel genetic algorithms. They distinguish three main categories of parallel genetic algorithms:

- Parallelization form master-slave on a single population
- Parallelization Fine-grained on a single population (diffusion model)
- Parallelization Coarse-grained on multiple populations (migration model)

In the first model, there is only one population residing on a single processor called the master. This one makes the different genetic operators of the algorithm on population and then distributes the evaluation of individuals to slave processors.

In the second model, which is suitable for massively parallel computers, the individuals in the population are distributed on processors, preferably at a rate of one individual per processor. Selection and reproduction of individuals operators are limited to their respective neighborhoods. However, as the neighborhoods overlap (an individual may be part of the vicinity of several other individuals), a certain degree of interaction between all individuals is possible.

The third category, more sophisticated and more popular, consists of several populations that are distributed over processors. These can evolve independently of each other with only occasional exchanges of individuals. This optional exchange called the migration phenomenon, is controlled by various parameters and generally provides a better performance of this algorithm type. This category is also called "parallel genetic algorithms islands".

2.1 Hybrid Parallelization of Metaheuristics

Each metaheuristic has its own characteristics and its own way to look for solutions. Therefore, it may be interesting to hybridize several different metaheuristics to create new research behaviors. In this regard, Bachelet et al. [10] identified three main forms of hybrid algorithms:

- Sequential hybrid, where two algorithms are executed one after the other, the results provided by the first being the initial solutions of the second.
- Synchronous parallel hybrid, where a search algorithm is used in place of an operator. An example of this type is to replace the mutation operator of genetic algorithm with a tabu search.
- Asynchronous parallel hybrid, where several search algorithms work concurrently and exchange informations.

2.2 Measuring Performance of Parallel Algorithms

In general, it's hard to make fair comparisons between algorithms such as meta-heuristics. The reason is that we can infer different conclusions from the same results depending on the metrics we use and how they are applied. This comparison become more complex when compared parallel metaheuristics, it's way is necessary to qualify some metrics, or even to adjust them to better compare parallel metaheuristics between them. Alba et al. [11] indicate that for non-deterministic algorithms, such as meta-heuristics, it is the average time of sequential and parallel versions which must be taken into account. It offers different definitions of speedup. Strong speedup which compares the parallel algorithm with the result of the best known sequential algorithm. This is what is closest to the true definition of speedup but considering the difficulty of finding each time the best existing algorithm, this standard is not used much. Speedup is called weak if we compare the parallel algorithm with the sequential version developed by the same researcher. It can then present its progress both in terms of quality and in pure speedup. Barr and Hickman [12] presented a different taxonomy consisting of relative speedup and absolute speedup. The relative speedup is the ratio between the parallel version running on a single processor and that performed on the set of processors. Finally, the absolute speedup, which is the ratio of the fastest sequential version on any machine and the execution time of the parallel version.

Speedup. The first and probably most important performance measure of a parallel algorithm is the speedup [11]. It is the ratio of the execution time of the best algorithm known on 1 processor and that of the parallel version. Its general formula is:

$$Speedup = \frac{Sequential\ execution\ time}{Parallel\ execution\ time}$$

Efficiency. Another popular metric is efficiency. It gives an indication of the rate of use of the requested processors. Its value is comprised between 0 and 1 and it can be expressed as a percentage. The more the value of efficiency is close to 1, the better is the performance. Efficiency equal to 1 matches to a linear speedup. Its general formula is:

$$Efficiency = \frac{Speedup}{P}$$

(P is the number of processors)

Other measures. Among other metrics used to measure the performance of parallel algorithms, we find the "scaled speedup" (expandable speedup) [11] which measures the use of available memory. We also find the "scaleup" (scalability) [11] to measure the ability of the program to increase its performance when the number of processors increases.

2.3   Impact of Communication on the Performance of Parallel Algorithms

The measure of parallel performance is a complex metric. This is mainly due to the fact that the parallel performance factors are dynamic and distributed. [13] The communication factor is among the most influential on the performance of the algorithm. In many parallel programs, the tasks performed by different processors need access to shared data. This creates a need for communication and slows the performance of the algorithm. These communications are more important in the case where processors require data generated by other processors. These communications are minimized in terms of data volume and frequency of exchanges when we used our new replacement approach, which is a hybrid approach that uses both Lamarckian and Baldwinian approaches at the same time, and this is the object of the next section.

2.4   Lamarckianism vs. Baldwinian Effect

When integrating local search with genetic algorithm we are faced with the dilemma of what to do with the improved solution that is produced by the local search. That is, suppose that individual i belongs to the population P in generation t and that the fitness of i is f(i). Furthermore, suppose that the local search produces a new individual i' with f(i') < f(i) for a minimisation problem. The designer of the algorithm must now choose between two alternative options. Either (option 1) he replaces i with i', in which case P = P −{i} + {i'} and the genetic information in i is lost and replaced with that of i', or (option 2) the genetic information of i is kept but its fitness altered: f(i) = f(i'). The first option is commonly known as Lamarckian learning while the second option is referred to as Baldwinian learning. The issue of whether natural evolution was Lamarckian or Baldwinian was hotly debated in the nineteenth century until Baldwin suggested a very plausible mechanism whereby evolutionary progress can be guided towards favorable adaptation without the inheritance of life-time acquired features. Unlike in natural systems, the designer of a Memetic Algorithm may want to use either of these adaptation mechanisms. Baldwin effect could be used to improve the evolution of artificial neural networks, and a number of researchers have studied the relative benefits of Baldwinian versus Lamarckian algorithms. Most recent work, however, favored either a fully Lamarckian approach, or a stochastic combination of the two methods. It is a priori difficult to decide what method is best, and probably no one is better in all cases. Lamarckianism tends to substantially accelerate the evolutionary process with the caveat that it often results in premature convergence. On the other hand, Baldwinian learning is more unlikely to bring a diversity crisis within the population but it tends to be much slower than Lamarckianism.

In our PMA, in each slave machine, when the Tabu Search algorithm runs on individuals sent by the master machine, and before returning improved individuals, we have to decide which replacement strategies will be applied. This decision will be taken according to the fitness value of the improved individual. When this fitness is lower than predefined threshold, we don't need to the genetic information of the individual, but we have to send his fitness to the master, in this case, we will send just the fitness value of the individual without its genetic information to the master to replace it in population with the Baldwinian approach, otherwise if the fitness value of the improved individual is above then the predefined threshold, in this case, we need to send the genetic information and the fitness value of the individual to the master to replace it in population with the Lamarckian approach.

3   Adaptive Memetic Algorithm

We present the adaptation of the Memetic Algorithm (MA) [14, 15] for the Classification problem. In the literature, there are two different approaches to extract rules using a genetic algorithm: the Pittsburgh approach and the Michigan one [15]. In our work we have chosen the Michigan approach where a classification rule presents the following form:

$$A \longrightarrow C$$

A is the premise or antecedent of the rule and C the predicted class. The A part of the rule is a conjunction of terms that are of the form:

Attribue	Operator	Value

The rule coding involves a sequence of genes arranged in the same order as the attributes of the studied data except for the last gene of the individual or chromosome which contains the predicted value of class [16]. Each condition is coded by a genome and consists of a triplet of the form (Ai op Vij), where Ai is the ith table attribute on which the algorithm is applied. The term op is one of the operators ' = ', ' < ' or ' > ' and Vij is the Ai attribute value belonging to its values domain. To each genome is associated a boolean field that indicates whether the premise is activated or not, in order to maintain the chromosome size fixed. Even if individuals have the same length, the

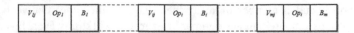

Fig. 1. Structure of an individual

rules associated with them are of variable length. The structure of an individual is shown in Fig. 1, where m is the total number of attributes.

The initial population is randomly generated to give it some diversity. Each individual (or rule) is a potential solution to the problem to solve. However, these solution do not all have same relevance degree. The rule coding involves a sequence of genes arranged in the same order as the attributes of the studied data except for the last gene of the individual of chromosome which contains the predicted value of the class. This is why the following criteria have been chosen [16]:

- To maximize the rule converge;
- To maximize the accuracy rate of the rule;
- To minimize the rule size because the comprehensibility of the rule is measured by the number of premises;

Fitness =
Λ_1 * Coverage/Total number of instance
+ Λ_2 * TP/Coverage
− Λ_3 * Rule size/Total number of attributes
where Λ_i is a real value that verifies $\sum \Lambda_i = 1$

In our Memetic Algorithm, we used hybridization of the tabu search with a genetic algorithm. we used the tournament selection and the classical genetic crossover and mutation operators. The individual resulting from crossover and mutation operators is the initial solution (a rule) for the tabu search, then the best individual found by the tabu search will replace the worst individual in term of accuracy in the population of the genetic algorithm and so on.

In the tabu search approach, the neighborhood of the initial solution consists of all solutions obtained by performing a one-movement operator which is applied to the current individual as many times as the number of attributes of the considered training set. So the created neighbors are evaluated by computing the same fitness as in the genetic algorithm. Then the best solution in the vicinity of the current individual is added to tabu list. Thus, the worst individual in term of accuracy is destroyed if the size of tabu list is exceeded and so on.

4   The Proposed PMA Architecture

We present in this section the design of our synchronous parallel Memetic Algorithm (PMA). It is a synchronous parallel model based on master-slave form uses a unique population residing on a single processor called the master. The latter performs the different genetic operations of the algorithm and then distributes the Tabu Search on the slave processors.

4.1 Replacement Strategy Used

In our PMA we hybridized the Lamarckian and Baldwinian approaches together to create a new approach in order to reduce the genetic information exchanged between the Genetic algorithm and the Tabu Search algorithm without penalizing the accuracy of the classifier based on our PMA. This hybrid approach is defined as follows:

- If the local search produces an individual i' with $f(i') >$ Threshold, in this case the Lamarckian approach is used, therefore $P = P - (i) + (i')$ and $f(i) = f(i')$
- If the local search produces an individual i' with $f(i') <=$ Threshold, in this case, the Baldwinian approach is used, therefore, P still the same and $f(i) = f(i')$

The Threshold is a variable parameter, its value determines the number of individuals which will be replaced with the Lamarckian approach, and the number of individuals that will be replaced with the Baldwinian approach.

4.2 Our Synchronous PMA Using Master-Slave Model

In this model, we have a master machine and the others are slave machines. In each slave machine, the Tabu Search algorithm runs on individuals sent by the master machine and before returning improved individuals we compare the fitness value of each individual to a predefined threshold. If the fitness value of the individual is more than the threshold then the genetic information and the fitness value of the individual are both sent to the master, otherwise, if the fitness value of the individual is lower than the threshold, we will send just the fitness value of the individual without its genetic information.

The memetic algorithm runs in the master processor, and the master is the only machine that has the overall population in its own memory. The master processor performs the selection, the crossover and the mutation of individuals and then distribute them to the slaves. Each slave processor receives the individuals, performs the tabu search and returns the optimized individuals to the master. When the master processor receives all results from slaves, he performed the replacement operation. If the master processor receives the fitness value of the an improved individual with his genetic information, then he replaced it in a population with the Lamarckian approach, else if he receives the fitness value of the improved individual without his genetic information, then he replaced it in a population with the Baldwinian approach.

The learning database is the only common data between the master and slaves. Consequently, we find the same learning database in all slaves. In order to always have the same learning database anywhere, the master machine sends the best individual selected after each generation of the Memetic Algorithm to all slave machines, for that they can update their learning database.

Master/Slave communication. The different types of communication can be summarized as follows:

From master to slave. The different informations sent from the master to a slave are:

- The individual resulting from the selection, crossover and mutation operators;
- The threshold value after each iteration;
- The best individual of each generation (Fig. 2);

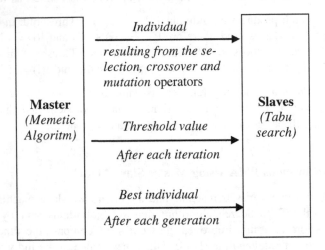

Fig. 2. Communication from master to slave

From slave to master. The different informations sent from a slave to the master are:

- The fitness value of the improved individual with its genetic information.
- The fitness value of the improved individual without its genetic information (Fig. 3).

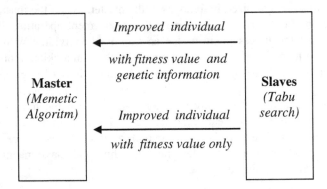

Fig. 3. Communication from slave to master

Synchronization. In this model the synchronization is launching of different slave processors. At first, the master launches them all, then each time the master needs to perform the Tabu Search on a set of the individuals distributes them on slave processors and waits for all results, then it replaces them in the population.

Slaves algorithm.

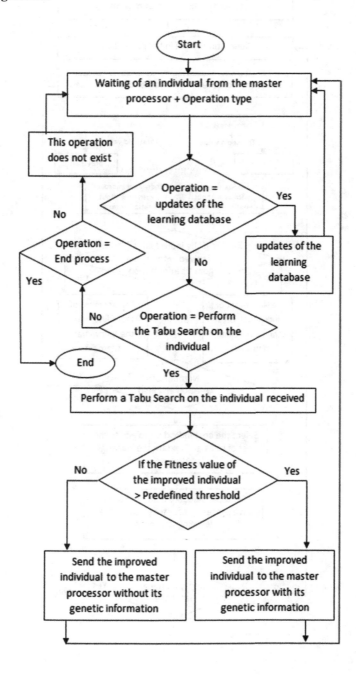

Master algorithm.

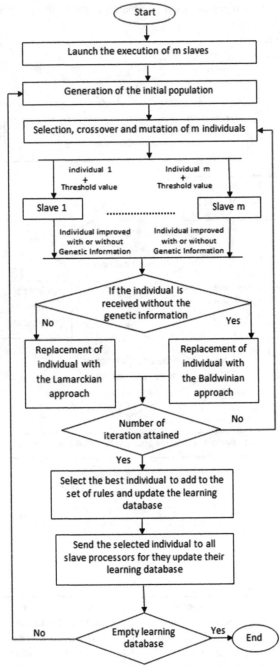

5 Results

5.1 UCI Benchmarks

The UCI is a very large database library of Benchmarks selected by the University of California Irvine (UCI) [17]. The latter was made available to the research community in Data Mining. These benchmarks that are widely used and considered as a reference. Hence the importance of using them to evaluate algorithms and compare their performance with other algorithms. Table 1 gives a summary of the databases used in our tests.

Table 1. Databases used

DataBases	Instances number	Attributes number	Class number
hepatitis	155	20	2
heart-statlog	270	14	2
segment-challenge	1500	20	7
ionosphere	351	35	2
kdd-train	11419	42	2
diabetes	768	9	2

5.2 Results Obtained by Our Synchronous PMA

The efficiency of our synchronous PMA is determined by the threshold value. In order to find the best threshold value, we conducted a series of experiments with three different thresholds:

- **Threshold1:** Is equal to the worst fitness value in the population for each iteration;
- **Threshold2:** Is equal to the best fitness value in the population for each iteration;
- **Threshold3:** Is equal to the average of all fitness values of the population for each iteration;

We run the classifier based on our synchronous PMA 10 times successively on the UCI benchmarks given above, for each threshold. We give every time the accuracy obtained, and the percentage of individuals returned without the genetic information compared to individuals returned with the genetic information.

The parameters PC and PM of Memetic Algorithm are PC = 0.025 and PM = 0.8 and the parameters $\lambda 1$, $\lambda 2$ and $\lambda 3$ of the objective function of classifier are $\lambda 1 = 0.1$, $\lambda 2 = 0.8$ and $\lambda 3 = 0.1$ and the parameters memory size and the number of iterations of the Tabu search are 5 and 300.

We have regrouped the average accuracy and the average percentage of individuals exchanged without their genetic information, obtained from all databases for each threshold in the following tables:

We observe from Tables 2 and 3, that the percentage of individuals returned without their genetic information for the threshold1 is between 0.08% and 0.36% maximum, so most individuals are returned with their genetic information and are

Table 2. Averages accuracies

Database	Average accuracy (%)		
	Threshold 1	Threshold 2	Threshold 3
hepatitis	84,57	73,30	84,39
heart-statlog	83,79	72,57	83,62
segment-challenge	94,81	83,39	96,06
ionosphere	90,35	87,24	90,14
kdd-train	99,29	86,62	99,79
diabetes	81,46	70,57	81,31

Table 3. Percentage of individuals returned without their genetic information

Database	Average percentage of individuals returned without their genetic information (%)		
	Threshold 1	Threshold 2	Threshold 2
hepatitis	0,28	99,84	41,98
heart-statlog	0,16	99,90	40,42
segment-challenge	0.08	99,42	42,36
ionosphere	0,28	99,74	44,58
kdd-train	0,36	99,84	42,48
diabetes	0,06	99,62	43,90

replaced with the Lamarckian approach in the population. So with the threshold1, our hybrid approach converges to the Lamarckian approach and we could not reduce the genetic information exchanged between master and slaves. On the other hand, the percentage of individuals returned without their genetic information for the threshold2 is between 99.42% and 99.90%, so most individuals are returned without their genetic information and are replaced with the Baldwinian approach in the population. So with the threshold2, our hybrid approach converges to the Baldwinian approach and we reduced by 50% the genetic information exchanged between the master and his slaves, but on the other hand we obtained bad results in the accuracy of the classifier, for this threshold. For the threshold3 the percentage of individuals returned without their genetic information is between 40.42% and 44.58%, so almost half of the individuals are returned without their genetic information, and also we have obtained very good results in terms of accuracy of the classifier. So with the PMA based on the threshold3, we could decrease by 20% the genetic information exchanged between the master and his slaves without penalizing the accuracy of the classifier.

Furthermore to evaluate the performance of our PMA designed, we have performed a series of tests on a network of 10 computers. The speedup, defined as the quotient between the time Ts to run the sequential algorithm and the time Tp for the parallel version, is used as the performance criterion.

To test the speedup of our PMA based on the threshold 3, we'll run it on the two databases hepatitis (20 attributes) and kdd-train (42 attributes) and each time we increase the number of slave processors, then we will compare it with the results of another simple PMA(is PMA without the new approach of replacement). The results found are in the following table:

Table 4. Results obtained with a different number of slaves

Number of slave processors	Speedup			
	hepatitis		kdd-train	
	Simple PMA	PMA with threshold3	Simple PMA	PMA with threshold3
1	1,00	1,00	1,00	1,00
2	1,84	1,93	1,65	1,73
3	2,85	2,99	2,56	2,69
4	4,17	4,37	3,75	3,94
5	5,18	5,69	4,14	4,76
6	5,47	6,01	4,37	5,03
7	6,06	6,66	4,52	5,42
8	6,46	7,42	4,68	5,81
9	6,69	7,69	4,80	5,95
10	6,77	7,78	4,84	6,12

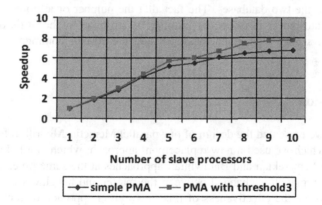

Fig. 4. Speedup of the two algorithms for hepatitis database

From the viewpoint of speedup, we observe from Table 4 and both Figs. 4 and 5 that the simple PMA and the PMA with threshold3 give both good results, every time we increase the number of slave processors, the speedup also increases. But if we compare the speedup of the two algorithms, we observe that they have almost the same speedup when the number of slaves is between 1 and 4, but once the number of slaves exceeds 4 the speedup of PMA with threshold3 becomes better than that of simple

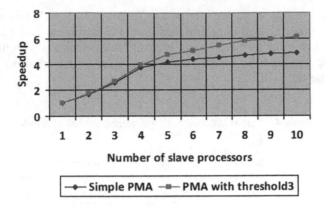

Fig. 5. Speedup of the two algorithms for kdd-train database

PMA for both databases hepatitis and kdd-train, which can be justified by the increase in the cost of communication between the slaves and the master for the simple PMA, the fact that the number of slaves increases the size of exchanged messages also increases, therefore, the communication costs slow the speedup. On the other hand the speedup of the PMA with threshold3 is better because the size of the messages exchanged is reduced by 20%, therefore, the cost of communications is reduced too, and the speedup is increased.

We also observe that the speedup of the two algorithms for hepatitis database is better than their speedup for kdd-train database, which can be justified by the number of attributes of the two databases. The fact that the number of attributes of kdd-train database (42 attributes) is twice the number of attributes of hepatitis database (20 attributes), the size of exchanged messages and the cost of communications is also double.

6   Conclusion

In this work, we presented the design of our parallel Memetic Algorithm for building a classifier. In which we used a new replacement approach, which is a hybrid approach that uses both Lamarckian and Baldwinian approaches at the same time, to reduce the quantity of information exchanged between the master and his slaves.

In order to see the effectiveness of this new hybrid approach of replacement and their effect on the quantity of information exchanged and on the accuracy of the classifier, we performed a series of tests on the UCI Benchmarks, and through the tests, it was found that we have decreased by 20% the quantity of information exchanged between the master and his slaves without penalizing the accuracy of the classifier.

To show the performance of our parallel Memetic Algorithm, we performed a series of tests on a network of 10 computers. Then we compared the speedup obtained by our parallel Memetic Algorithm with the speedup of another simple parallel Memetic Algorithm. It was observed that once the number of slaves exceeds 4, the speedup of our parallel algorithm is better than the simple parallel algorithm, because the number of messages exchanged in our parallel algorithm decreased by 20%, therefore the communication costs are reduced and the speedup is increased. It was also noted that if the number of attributes in the database used increases, therefore, the size of exchanged messages and the cost of communications also increases, hence the importance of our work in minimizing the quantity of exchanged individuals.

References

1. Cios, K.J., Pedryecz, W., Swinniarsky, R.W., Kurgan, A., et al.: Data Mining: A Knowledge Discovery Approach. Editions Springer Science (2007)
2. Jain, A.K., Dubes, R.C.: Algorithms for clustering data. Editions Prentice Hall Advanced Reference Series. Prentice Hall, New Jersey (1988)
3. Dréo, J., Pétrowski, A., Siarry, P., Taillard, E.: Métaheuristiques pour l'optimisation difficile. Eyrolles (2005)
4. Blum, C., Roli, A.: Metaheuristics in combinatorial optimization: overview and conceptual comparison. ACM Comput. Surv. **35**(3), 268–308 (2003)
5. Hao, J.-K., Galinier, P., Habib, M.: Métaheuristiques pour l'optimisation combinatoire et l'affectation sous contraintes. Revue d'intelligence artificielle (1999)
6. Goldberg, D.E.: Genetic Algorithms in Search, Optimization and Machine Learning. Addison Wesley, Massachusetts (1989)
7. Glover, F.: Tabu search - Part I. ORSA J. Comput. **1**(3), 190–206 (1989)
8. Crainic, T.G., Toulouse, M., et al.: Parallel Metaheuristics. In: Crainic, T.G., Laporte, G. (eds.) Fleet Management and Logistics, pp. 205–251. Kluwer Academic, Norwell (1998)
9. Cantú-Paz, E.: A survey of parallel genetic algorithms. Calculateurs parallèles, réseaux et systèmes répartis **10**(2), 141–171 (1998)
10. Bachelet, V., Hafidi, Z., Preux, P., Talbi, E.-G.: Vers la coopération des métaheuristiques. Calculateurs parallèles, réseaux et systèmes répartis 10(2) (1998)
11. Alba, E., Luque, G.: IV leasuring the performance of parallel metaheuristics. In: Parallel Metaheuristics: A new Class of Algorithms. Wiley-Interscience (2005)
12. Barr, R., Hickman, B.: Reporting Computational Experiments with ParaUel Algorithms: Issues, Measures, and Experts' Opinions. Dept. of Computer Science and Engineering, Southern Tvlethodist University (1992)
13. Malony, A.: Tools for Parallel Computing: A Performance Evaluation Perspective, ch. VII, p. 342. Springer (2000)
14. Bacardit, J.: Pittsburgh Genetic-Based Machine Learning in the Data Mining era: Representations, Generalization, and Run-time. Ph.d. Thesis, Universitat Ramon Llul, Spain (2004)
15. Witten, I.H.: Data Mining: Practical Machine Learning Tools and Techniques with JAVA Implementations. Morgan Kaufman Publishers, San Mateo (2003)
16. Tan, K.C., Yu, Q., Ang, J.H.: A dual-objective evolutionary algorithm for rules extraction in data mining. Comput. Optim. Appl. **34**, 273–294 (2006)
17. Blake, C.L., Merz, C.J.: UCI repository of machine learning databases (1998)
18. Moscato, P.: On evolution, search, optimization, genetic algorithms and martial arts: Towards memetic algorithms. Caltech concurrent computation program, C3P Report 826 (1989)

Classical Mechanics Optimization for Image Segmentation

Charaf Eddine Khamoudj[1]([⊠]), Karima Benatchba[1],
and Mohand Tahar Kechadi[2]

[1] Laboratoire des Méthodes de Conception de Systèmes, Ecole nationale
Supérieure d'Informatique, Oued Smar, Algiers, Algeria
{c_khamouj,k_benatchba}@esi.dz
[2] School of Computer Science, University College Dublin, Dublin, Ireland
tahar.kechadi@ucd.ie

Abstract. In this work, we focus on image segmentation by simulating the natural phenomenon of the bodies moving through space. For this, a subset of image pixels is regularly selected as planets and the rest as satellites. The attraction force is defined by Newton's third law (gravitational interaction) according to the distance and color similarity. In the first phase of the algorithm, we seek an equilibrium state of the earth-moon system in order to achieve the second phase, in which we search an equilibrium state of the earth-apple system. As a result of these two phases, bodies in space are constructed; they represent segments in the image. The objective of this simulation is to find and then extract the multiple segments from an image.

Keywords: Image segmentation · Combinatorial optimization · Artificial intelligence · Metaheuristic · Classical mechanics optimization

1 Introduction

Segmentation is an important step in the image processing; it extracts segments from images. Each segment represents a set of pixels (each pixel is defined by its coordinates and color).

Image segmentation can be seen as a combinatory optimization problem, because the goal is to find combinations of assigning pixels to segments. To find optimal partitioning in *K* groups of an *n* pixels image, all the possible partitions must be browsed. The number of possible partitions is given by the Stirling numbers of the second kind [1]:

$$S(n,k) = \frac{1}{k!} \sum_{i=1}^{k} (-1)^{k-i} \binom{k}{i} i^n \quad Where : \binom{k}{i} = \frac{k!}{i!(k-i)!} \tag{1}$$

If the optimal number of partitions is unknown, Stirling numbers are calculated for *k = 1* to *k = n*. The number of possible partitions is given by Bell number [1]:

© Springer International Publishing AG 2016
P. Siarry et al. (Eds.): ICSIBO 2016, LNCS 10103, pp. 102–110, 2016.
DOI: 10.1007/978-3-319-50307-3_8

$$B(n) = \sum_{k=1}^{n} S(n, k) \qquad (2)$$

The Bell number quickly becomes very big (example: $B(10) = 115975$). The heuristic approaches for solving a combinatorial problem is to find a good solution in a bounded time among an exponential number of possibilities. So they are based on finding a good compromise between the calculation time and the quality of the best solution found so far.

The objective is to use the state of bodies' equilibrium in the space as a heuristic to tackle the image segmentation problem. We have proposed and implemented an image segmentation method based on a new metaheuristic inspired by the natural phenomenon of the bodies' movement in space. The proposed metaheuristic is based on the impact of the attractive forces between the bodies during their movements.

To simulate this problem as a natural phenomenon of the bodies' movement in space, we need to define the planets, the satellites, and what the attraction force. For this, m pixels of the image are uniformly selected. These pixels represent the planets, the remaining pixels represent the satellites and the attraction force is defined by the color similarity and the distance between the planet and the satellite.

The earth-moon system equilibrium is to find a situation, in which every single satellite is in rotation over the planet that applies on it the strongest attraction force. The earth-apple system equilibrium is to find a situation, in which all the bodies are far from colliding on each other, the resulting bodies represent the segments of an image.

2   Metaheuristics Inspired from the Interaction Force

In the universe, attraction forces are divided into two types: Gravity is an attractive force between the bodies, which depends on their masses. The electromagnetic interaction is an attractive force that acts on the elements with electrical charges. Some researchers have proposed metaheuristics based on the forces of attraction between bodies. These forces are generated either from the physical mass or the electric charge. Here are some examples of this type of metaheuristic:

2.1   Gravitational Search Algorithm (GSA)

The gravitational search algorithm [2] uses Newton's third law to calculate the forces of attraction and Newton's second law to deduce the speed of a body. The diversification of the search in ensured by attraction force; To intensify the search, the gravitational constant is linearly decreased with time. GSA algorithm is combined with Particle Swarm Optimization (PSO) to solve the image segmentation problem [3]; The result algorithm is used in the second phase to search for the optimal threshold estimation used as a search procedure in the first phase.

2.2   Charge Search System (CSS)

The search system based on the electric charge [4] is inspired by the electrostatic; attributing electrical charges to the particles. The algorithm is used as a step of local search to improve the founded solutions in PSO algorithm [5] to solve the image segmentation problem.

2.3   Gravitational Interactions Optimization (GIO)

Optimization by gravitational interactions [6] called particle swarm optimization with gravitational interactions. Each body stores its current position and its best position. The interactions of bodies follow the Newton's third law and move each body to a new location so that the whole population tends to reach the optimum. This method uses the Newton's second law to calculate the speed of a body. To intensify the search, authors use a mass unit placed in space to exert forces on other bodies to move them. When the bodies are close to each other, the resulting forces are strong, and there are many displacements.

2.4   Fusion-Fission Metaheuristic

The fusion-fission metaheuristic [7] is inspired from nuclear physics. It is applied on the graph partitioning problem, the clustering of documents and image segmentation. The atom is formed of electrons with a negative charge and nucleons which form the atomic core. There are two kinds of Nucleons: protons, positively charged and neutrons, neutrally charged. The cohesion of the atomic core is ensured by their strong interactions. During the fission of an atom, the core divides into two fragments, along with several ejected neutrons. An atom can split either spontaneously if its core is too heavy, or because of being hit by a neutron. To merge, atoms must have sufficiently high speeds. He considers a cloud of nucleons. It is subjected to high temperature and pressure, so that the nucleons have great chances of collision. It is the fusion of these nucleons together that forms the resulting atoms, which will help achieve an equilibrium state of the system. Fission is used to explode the biggest or non stable atoms.

3   Classical Mechanics Optimization (CMO)

As mentioned earlier, metaheuristics based on the gravitational interaction are hybridized with other metaheuristics, such as GSA algorithm with simulated annealing, and GIO algorithm that is hybridized with the particle swarm optimization. The proposed method is independent; it relies on applying the laws of classical mechanics.

The CMO simulates the natural phenomenon of the bodies' movement in a space by considering the pixels as bodies. m of these pixels are selected as planets and the n remaining are considered satellites, the attraction force is defined by Newton's third law (gravitational interaction).

After the simulation of the problem as a system of bodies in space, we execute the algorithm in two main phases: The first phase is to find an equilibrium of the rotating satellites around planets by applying the earth-moon system. The second phase is to group the segments formed in the first phase by applying the earth-apple system.

3.1   Transformation of the Problem into a System of Bodies in Space

The planets represent a subset S of the set E (E is the global set that contain all pixels of image), and satellites represent the subset N representing the complement of S in E. Rules (3) and (4) are to calculate the number of planets and the number of satellites:

$$m = PixelNbr \times \frac{PlanetNbr}{PlanetNbr + SatelliteNbr} \tag{3}$$

$$n = PixelNbr - m \tag{4}$$

The following figure shows the image to segment, the black pixels represent the planets, the remaining are satellites (Fig. 1).

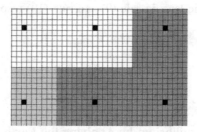

Fig. 1. Planets selection.

The number of pixels of each segment defines their mass:

$$ClassMass = UnitMass \times ClassPixel\ Nbr$$
$$Where:\ UnitMass = \frac{SystemMass}{PixelNbr} \tag{5}$$

Newton's third law (gravitational interaction) is used to define the attraction force.

$$F_{ab} = g\frac{m_a m_b}{d_{ab}^2} \tag{6}$$

Where m_a, m_b represent masses and d_{ab} represents distance between pixels a and b.

The distance d_{ab} is calculated by Euclidean distance, after the simulation of the spatial distance from a triangular rule:

$$GreatestPixelDistance(GPD) \rightarrow GreatestBodiesDistance(GBD)$$
$$PixelDistance(PD) \rightarrow BodiesDistance(BD)$$

So the distance ratio becomes:

$$\text{Distance Ratio} = GBD/GPD \tag{7}$$

The image is composed by a matrix of pixels; each pixel is defined by its coordinates and its color. After experimentation, the equation of attraction is improved as follows:

$$F_{ab} = \frac{m_a m_b}{d_{ab} e^{\sqrt{|c_a - c_b|}}} \tag{8}$$

Where c_a and c_b represent the color of the pixel a and the pixel b.

The gravitational fields earth-moon system GF_{em} and earth-apple system GF_{ea} are derived from the mechanic laws in rule (9) and rule (10) respectively:

$$GF_{em} = 1.068 \times mass \times 10^{-21} \tag{9}$$

$$GF_{ea} = GF_{em}/200 \tag{10}$$

3.2   Finding a Body Equilibrium by Applying the Earth-Moon System

We look in space for an equilibrium of the bodies, to stabilize the movement of satellites around planets. The movement of the satellites is caused by the gravitational attraction exerted by the planets.

A body a is rotating around the body b with a force F_{ab}. If there is a body c where: $F_{ac} > F_{ab}$, then the body a leave its path around b and follows a new path around c.

For each combination, we calculate the attraction force for planets. The gravity center becomes the center of all satellites around this planet.

We repeat the two previous steps until the system equilibrium is verified. The algorithm of this step is described as follows:

The following figure shows a stable distribution of the satellites around the planets (Fig. 2).

```
Algorithm 1 Find the system equilibrium {Earth-Moon sys-
tem}
   var    E: array [1..z,1..w] of real; {E is the image
             where z and w are the dimensions}
          Planet: array [1..m,1..n] of integer; {Each
             row of this matrix represents the satellites
             turned around the corresponding planet Since
             Planet(j, 1) represents the center of
             gravity of the same line}
          Satellite: array [1..n] of integer; {Each case i
             represents the corresponding planet of
             satellite i}
          m, n : integer; {m is planet number and n is
             satellite number }

  begin
    Calculate(m); Calculate(n);
    Initialize(Planet, Satellite);
    repeat
      For i = 1 to n
        For j = 1 to m
          If Earth-Moon gravitational field (Planet(j))
               > Distance (Planet(j),Satellite(i)) then
            If Force (Planet (j,1), Satellite(i)) > Force
               (Planet (Satellite (i),1),Satellite (i))
            then
               Move (Satellite(i), j);{Is to release the
                  satellite i from his planet and assign it
                  to the planet j}
             end;
          end;
        end;
      end;
    Until system stabilization
end.
```

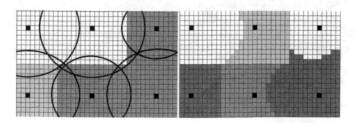

Fig. 2. System equilibrium for earth-moon system and grouping of bodies.

3.3   Construction of Segments by Applying the Earth-Apple System

After the stabilization of satellites around planets, each planet-moon system is considered as one body. Then the gravitational fields of bodies (earth-apple system) is calculated. Each body situated in the gravitational field of another body falls (fusion of two segments) and the two bodies are considered as a single body. After the fusion, we repeat the previous two steps until the overall system is stable (all the found segments are too far to be fused). The algorithm of this phase is as follows:

```
Algorithm 2 Research earth-apple system equilibrium
  use Result of Algorithm 1
  begin
    repeat
      For i = 1 to n
        For j = 1 to m
          If Gravitational field earth-apple(Planet(i))
              > Distance(Planet(i), Planet(j)) then
            Move (Planet(j), i); {move all pixels of the
              body j to the body i, recalculate the new
              center of gravity and  remove the line j
              from the Planet Matrix}
            m := m-1;
          end;
        end;
      end;
    Until system stabilization
  end.
```

The following figure shows the bodies gravitational fields and fusion (earth-apple system) (Fig. 3).

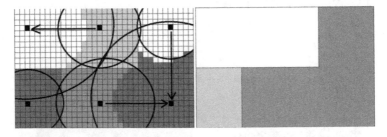

Fig. 3. Bodies gravitational fields and fusion (earth-apple system).

3.4 Tests and Results

We applied the CMO approach on real images by simulating the solar system. The number of pixels planets m is calculated as follows:

$$m = \text{the number of pixels in the image} * 8/(167 + 8)$$

Because in the solar system there are one hundred and sixty seven (167) satellites and eight (08) planets. The results of the segmentation are:

The segmentation of the two images provides three segments that represent the objects of each image, that the dice coefficient of the approach is 88,65. There are small segments that are not displayed, it is the influence of light or stains. This is positive because these pixels or smaller segments can be treated isolated segments which are used to solve other problems such as the detection of tumors in medical imaging (Fig. 4).

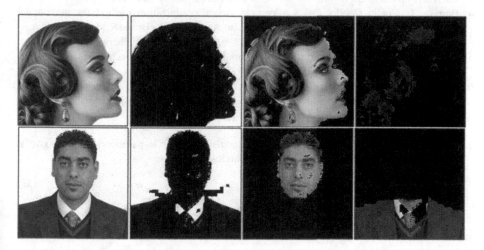

Fig. 4. Image segmentation results by using CMO.

4 Conclusion

In this work, we developed an image segmentation method based on the simulation of the natural phenomenon of bodies' movement in space, called Classical Mechanics Optimization. It consisted on two phases. As a first step, we seek an equilibrium of the bodies in the earth-moon system (satellites assignment to the planets that apply more attraction force). The second step is to group the most similar segments, applying an earth-apple system.

The simulation is made by the extraction of a pixels subset as planets and the remaining pixels represent the satellites. The attraction force equation is defined by the rule (8) that represents Newton's third law according to the pixels' color, considering the importance of color in image segmentation.

In CMO, intensification and diversification are provided by the distance ratio that transforms the distance between pixels in the spatial distance. When it is small, the algorithm becomes more intensive because the gravitational field increases. However, if the distance ratio is very small, all the bodies may fall into a black hole which represents a segment that includes all pixels. Otherwise, if it is very large, grouping objects is not assured, because of the decrease in the gravitational field.

References

1. Benzaghou, B., Barsky, D.: Nombres de Bell et somme de factorielles. Journal de Théorie des Nombres de Bordeaux **16**, 1–17 (2004)
2. Rashedi, E., Nezamabadi, H., Saryazdi, S.: Gsa. A gravitational search algorithm. Information Sciences, 179(13): 2232–2248 (2009)
3. Amandeep, K., Charanjit, S., Amandeep, S.B.: SAR image segmentation based on hybrid PSOGSA optimisation algorithm, vol. 4, issue 9 (2014). ISSN 2248-9622
4. Barrera, J., Coello Coello, C.A.: A particle swarm optimization method for multimodal optimization based on electrostatic interaction. In: Aguirre, A.H., Borja, R.M., Garciá, C.A.R. (eds.) MICAI 2009. LNCS, vol. 5845, pp. 622–632. Springer, Heidelberg (2009)
5. Dahiya, A., Dubey, R.B.: Survey of some multilevel thresolding techniques for medical imaging, vol. 3 issue 7 (2015). ISSN 2347-3878
6. Flores, J.J., López, R., Barrera, J.: Particle swarm optimization with gravitational interactions for multimodal and unimodal problems. In: Sidorov, G., Hernández Aguirre, A., Reyes Garc \'ıa, C.A. (eds.) MICAI 2010, Part II. LNCS, vol. 6438, pp. 361–370. Springer, Heidelberg (2010)
7. Bichot, C.: Elaboration d'une nouvelle métaheuristique pour le partitionnement de graphe Doctoral thesis. The Polytechnic National Institute of Toulouse (2007)

On the Community Identification in Weighted Time-Varying Networks

Youcef Abdelsadek[1]([✉]), Kamel Chelghoum[1], Francine Herrmann[1],
Imed Kacem[1], and Benoît Otjacques[2]

[1] Laboratoire de Conception, Optimisation et Modélisation des Systèmes,
Université de Lorraine, Metz, France
{youcef.abdelsadek,kamel.chelghoum,francine.herrmann,
imed.kacem}@univ-lorraine.fr
[2] e-Science Research Unit, Environmental Research and Innovation Luxembourg,
Institute of Science and Technology, Belvaux, Luxembourg
benoit.otjacques@list.lu

Abstract. The community detection play an important role in understanding the information underlying to the graph structure, especially, when the graph structure or the weights between the linked entities change over time. In this paper, we propose an algorithm for the community identification in weighted and dynamic graphs, called *Dyci*. The latter takes advantage from the previously detected communities. Several changes' scenarios are considered like, node/edge addition, node/edge removing and edge weight update. The main idea of *Dyci* is to track whether a connected component of the weighted graph becomes weak over time, in order to merge it with the "dominant" neighbour community. In order to assess the quality of the returned community structure, we conduct a comparison with a genetic algorithm on real-world data of the ANR-Info-RSN project. The conducted comparison shows that *Dyci* provides a good trade-off between efficiency and consumed time.

Keywords: Dynamic networks · Community detection · Genetic algorithm · Weighted graphs · Twitter's networks

1 Introduction

With the popularization of social networks like Twitter, an exponential quantity of data is generated. These data are increasing each day, and the existing algorithms which are not considering the dynamic nature of data would suffer from the scalability issue. Graphs are well-known to be appropriate to represent relationships between entities. For example, we can cite social relationships like the friendship, the follow in social blogs or the information sharing in social media. Additionally, these relationships can change over time with, appearing and/or disappearing relationships for binary relationships (e.g., friendship) or increasing and/or decreasing for weighted relationships (e.g., the number of times where two persons share information). This evolving complex network is called dynamic

© Springer International Publishing AG 2016
P. Siarry et al. (Eds.): ICSIBO 2016, LNCS 10103, pp. 111–123, 2016.
DOI: 10.1007/978-3-319-50307-3_9

graph. In this paper, we consider edge weighted dynamic graphs, where a weight is assigned to each edge of the graph. Furthermore, the community detection in dynamic graphs enhances our understanding of the underlying semantic behind the graph. The changes that might occur can be, either structural, attributes (i.e., weights) or also both of them [1]. Consequently, how to analyse the evolution of the communities structure over time? To answer this question, one need to devise an algorithm which relies on the graph features and which takes advantage from the previously identified communities by avoiding the community identification from scratch at each instant. As a concrete example, an analyst needs to understand how the information is shared in Twitter by understanding the role of each Twitter user within its community and outside of its community. To fulfil this need, one have to detect the communities of the analyst's time point of interest and to follow the community's member evolution with new members joining/leaving the studied communities. In this context, a trade-off between efficiency and response time is necessary to detect the community evolution over time.

More formally, a dynamic graph of an initial graph G_0 can be seen as a sequence of static graphs [2], denoted by $G_s = (G_0, G_1, \dots, G_f)$ with f snapshots giving rise to $Cs_s = (Cs_0, Cs_1, \dots, Cs_f)$ community partitions as results of $U_s = (U_0, U_1, \dots, U_{f-1})$ updates as illustrated in Fig. 1. We denote by N_t, E_t, E_t^w, N_s and E_s respectively, the set of nodes of size v at instant t, the set of edges at instant t, the set of edge weights at instant t of G_t, the set of nodes of the whole G_s and the set of edges of the whole G_s. Furthermore, the set of updates U_t varies in terms of the impact they cause to the current set of communities. As an instance, the impact of adding a new node and those of updating the weight of an existing edge differ. We point out that weights can be assigned to the nodes also, with node weight update scenario for the dynamic context, which is not considered in this paper. The following updates cases describes the repercussion on N_t, E_t and E_t^w after each update scenario.

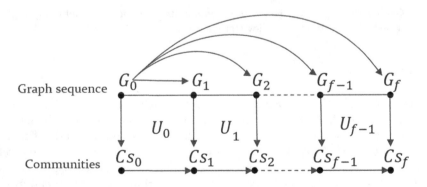

Fig. 1. The graph sequence of a dynamic graph

1. **Structural updates**:
   (a) **Node removing**: An old node on is removed, $N_{t+1} \leftarrow N_t \setminus \{on\}$ with the related edges.
   (b) **Edge removing**: An old edge oe is removed, $E_{t+1} \leftarrow E_t \setminus \{oe\}$.
   (c) **Node addition**: A new node nv is added, $N_{t+1} \leftarrow \{nv\} \cup N_t$ with the related edges.
   (d) **Edge addition**: A new edge ne is added, $E_{t+1} \leftarrow \{ne\} \cup E_t$.
2. **Attributes updates**:
   (a) **Edge weight updating**: An new edge weight ne^w of an old edge weight oe^w is updated, $E_{t+1}^w \leftarrow (E_t^w \setminus \{oe^w\}) \cup \{ne^w\}$.

The outlines of the remaining sections of this paper are as follow: In Sect. 2, the related works of the addressed topic are presented. Section 3 introduces the *Dyci* algorithm for community identification in weighted and dynamic graphs. Section 4 describes the genetic algorithm. In Sect. 5, the conducted assessments and the obtained results are discussed. Finally, Sect. 6 concludes this paper and gives the perspectives.

2   Related Work

This section presents some related works for the dynamic community detection. There are more algorithms for the static version of this problem in the literature compared to the dynamic case, especially, for those considering weighted edges. The static community detection algorithms can be divided into two families of algorithms: the divisive family (top-down) [3] and the agglomerative family (bottom-up) [4]. Concerning the dynamic community detection problem, in [5] it was proved to be *NP*-complete and *APX*-hard. For the unweighed dynamic community detection, the authors of [6] propose a matching algorithm to detect similar communities over the snapshots of the graph sequence, forming a meta-community which is the sequence of these identical communities. Agglomerative modularity-based approach are considered in [7–9]. The authors of [9] use a physical metaphor with forces, which retains a node to stay in its community against attracting forces of the other communities. Furthermore, game-theory analogy is used in [10]. In the latter, each node of the graph is considered as an agent, which maximizes its utility function. A set of predefined agent actions is set initially. The system ends when all agents choose their best community belonging (i.e., which maximizes the utility function). Relying on the colouring problem, a constant-approximation algorithm was proposed in [11]. The authors of [12] deal with changes tracking of communities in large networks. They propose an approach which uses agglomerative clustering to examine the evolution of the community structure over time by identifying stable communities after several cluster running. In [13] a model is described which tracks communities over time, those are characterised by a set of events. Regarding the weighted version of this problem, label propagation is used [14]. The idea of this algorithm is to allow a specific node to change its community label taking into account its adjacent nodes labels.

3 Dynamic Community Detection Algorithm

3.1 Notations and Definitions

Let us define c_{n_i}, IW, INW, WD and WI which, respectively, represent the community of the node n_i, the intra-community weight, the inter-community weight, the weighted degree of a node and the weighted community-incidence of a node. These are presented in the following equations:

$$IW_{c_g} = \sum_{n_i \in c_g} \sum_{n_j \in c_g} \frac{e^w_{n_i,n_j}}{2} \tag{1}$$

$$INW_{c_g,c_h} = \sum_{n_i \in c_g} \sum_{n_j \in c_h} e^w_{n_i,n_j} \tag{2}$$

$$WD_{n_i} = \sum_{j=1}^{v} e^w_{n_i,n_j} \tag{3}$$

$$WI(node, c_g) = \frac{\sum_{n_i \in c_g} e^w_{node,n_i}}{WD_{node}} \tag{4}$$

3.2 *Dyci* algorithm

First, an improved version of the algorithm proposed in [15] is applied on G_0 as a starting point Cs_0 of *Dyci*. The main idea of this algorithm consists in using a collection of pairwise node-disjoint triangles as a starting point to detect community structure of the graph. Then, adjacent communities are iteratively compared in terms of weights and merged when a merging condition holds. This iterative process ends when no community merging is observed. After, for each snapshot of the graph sequence *Dyci* reacts depending on the update scenario that occurs as presented in the Algorithm 1.

The following subsections show how *Dyci* reacts depending on the update scenario, each update case is considered and presented in detail.

Algorithm 1. *Dyci algorithm*

Input: Cs_t and U_t;
Output: Cs_{t+1}.
    BEGIN
        for each *oldNode* **in** $U_t.nodeToRemove$ **do**
            $Cs_t \leftarrow NodeRemoving\ (oldNode, Cs_t)$;
        end for
        for each *oldEdge* **in** $U_t.edgeToRemove$ **do**
            $Cs_t \leftarrow EdgeRemoving\ (oldEdge, Cs_t)$;
        end for
        for each *newNode* **in** $U_t.nodeToAdd$ **do**
            $Cs_t \leftarrow NodeAddition\ (newNode, Cs_t)$;
        end for

---

for each *newEdge* **in** $U_t.edgeToAdd$ **do**
   $Cs_t \leftarrow EdgeAddition~(newEdge, Cs_t)$;
end for
for each *edgeWeightUpdate* **in** $U_t.edgeWeightUpdate$ **do**
   $Cs_t \leftarrow EdgeWeightUpdating~(newEdgeWeight, Cs_t)$;
end for
Return Cs_t;
END.

3.3   NodeRemoving (*oldNode*):

The main idea of the node removing case is to check whether the deletion of *oldNode* generates several connected components or reduces the $IW(c_{oldNode})$. To this end, *Dyci* tests for each resulting connected component, noted CC, whether it can form a community by it self or would be merged with an adjacent community, noted *com*. In other words, *Dyci* verifies whether Eq. 5 holds or not. Figure 2 gives an example of the node removing update scenario.

$$INW_{com,CC} \geqslant IW_{CC} \tag{5}$$

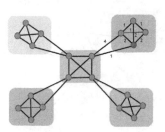   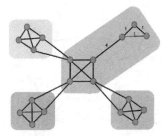

(a) Before removing the red node   (b) After removing the red node

Fig. 2. Node removing scenario example (Color figure online)

3.4   EdgeRemoving (*oldEdge*):

When an inter-community edge is removed, this reduces the inter-community weight leading to more community-like structure. However, the other case might lead to intra-community dividing in two connected components or a significant weight loss. To handle this case, *Dyci* compares weights between each resulting connected component of *oldEdge* deletion and their adjacent communities by Eq. 5. The edge removing update scenario is illustrated with an example in Fig. 3.

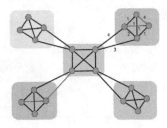
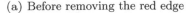

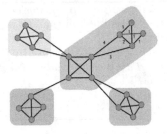

(a) Before removing the red edge   (b) After removing the red edge

Fig. 3. Edge removing scenario example (Color figure online)

3.5   NodeAddition (*newNode*):

Two subcases can occur for node addition. The first one is the subcase where
newNode has no community edge incidence leading to an isolated community. In
the second subcase, *newNode* comes with many edges. For the latter, *newNode*
is added to the community with the greatest $WI(newNode, c), \forall c \in$ communities
adjacent to *newNode*. Figure 4 gives an example of the node addition update
scenario.

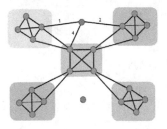

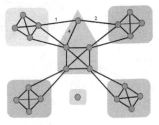

(a) Before adding the green nodes   (b) After adding the green nodes

Fig. 4. Node addition scenario example (Color figure online)

3.6   EdgeAddition (*newEdge*):

For this case, if *newEdge* is inserted inside a community, this will not affect the
community partition in terms of weights. Unlike, an inter-community edge could
increase the inter-community weight, noted c_1 and c_2, aggregating them in one
community. To handle this case, *Dyci* verifies whether Eq. 6 holds or not. The
edge addition update scenario is illustrated with an example in Fig. 5.

$$INW_{c_1,c_2} \geqslant IW_{c_1} \text{ or } INW_{c_1,c_2} \geqslant IW_{c_2} \tag{6}$$

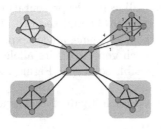 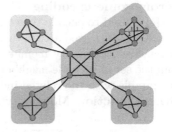

(a) Before adding the green edges   (b) After adding the green edges

Fig. 5. Edge addition scenario example (Color figure online)

3.7 EdgeWeightUpdating (*edgeWeightUpdate*):

The last case be partitioned into two subcases, illustrated in Fig. 6. The first subcase is when the *edgeWeightUpdate* is an inter-community edge with weight greater than the old edge weight. For this scenario *Dyci* verifies whether Eq. 6 holds or not. The second subcase rises when *edgeWeightUpdate* is an intra-community edge with weight lower than the old edge weight. The algorithm checks whether this weight loss leads to an adjacent community merging by Eq. 5.

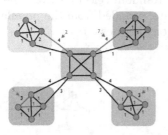 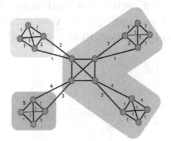

(a) Before updating the red edges weights

          (b) After updating the red edges weights

Fig. 6. Edge weight update scenario example (Color figure online)

4 Genetic Algorithm

Genetic algorithms (GA) can provide very good results if they are well set. In order to evaluate the quality of the obtained communities of Cs_f, a comparison is conducted between *Dyci* and the following GA.

- **Chromosome encoding**: The Locus-based Adjacency Representation [16] (LAR) is used to encode the community detection problem, like in [17,18]. In the LAR a $|N_f|$ sized array is used, where the couples (gene, allele) express an associative community membership. Indeed, each gene takes its allele value from the set of its node neighbours ensuring feasible solutions. Figure 7 shows an example with the related individual decoding.
- **Fitness function**: Modularity φ of [19] is used for individual evaluation:

$$\varphi = \frac{1}{2M} \sum_{n_i} \sum_{n_j} \left(e_{n_i,n_j}^w - \frac{WD_{n_i}WD_{n_j}}{2M} \right) \delta\left(c_{n_i}, c_{n_j} \right) \tag{7}$$

Where, $M = \sum_{i<j} e_{n_i,n_j}^w$ and $\delta\left(c_{n_i}, c_{n_j} \right) = 1$, if $c_{n_i} = c_{n_j}$, 0 otherwise.

The modularity expresses whether the detected community structure is well defined or not, corresponding to the density of the detected communities minus the density of these communities for the random case with the same characteristics.

- **Population initialization**: A random population of size 100 is generated and sorted in a decreasing fitness function order.
- **Crossover**: Uniform crossover with probability 0.9 is performed, as illustrated in Fig. 8a.
- **Mutation**: Random allele flipping with probability 0.1 is performed, as showed in Fig. 8b.
- **Parent selection and child insertion**: Random selection from the 20 % eliteness individuals. Weakest individuals are excluded from the population.
- **Stopping condition**: Number of generations reaches 50.

Fig. 7. An individual example using LAR encoding

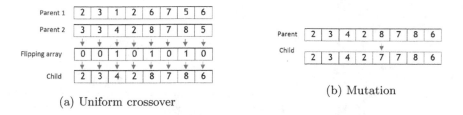

(a) Uniform crossover

(b) Mutation

Fig. 8. Reproduction operators

5   Experiments and Results Discussion

This section discusses the obtained results of the conducted comparison between the above GA and *Dyci* on four data sets from real-world data of the ANR-Info-RSN project. The ANR-Info-RSN project deals with the community detection in a collected set of tweets from social media. To this end, a graph is used as model leading to a weighted graph, where each Twitter's user of the collected data is represented by a node and an edge represents a retweet relationship between two Twitter's users. In this context, the edge weight is equal to the number of times where a retweet is observed between two Twitter's users. The experiments are conducted using our tool $NLCOMS$ [20]. Table 1 presents the data sets characteristics where the unit of snapshot generation is one day. Figures 9 and 10 show, respectively, the obtained results for the data sets at t_f and the averages values of the results for the whole Cs_s.

From Fig. 9a, we remark that *Dyci* and the GA have almost the same results (GA is slightly better than *Dyci*), taking into account the fact that the obtained communities Cs_f of *Dyci* are highly influenced by the f previous choices made during the whole graph sequence. One could say that *Dyci* obtains satisfactory results. Further, from Fig. 9c, we remark that *Dyci* is relatively fast compared to the GA, due to the fact that *Dyci* takes advantage from the previous identified community avoiding relaunching the process at each snapshot. From Fig. 10, we notice that the averages values are almost the same by comparing to the values of the last snapshot t_f, except for DS3 where *Dyci* takes more time and provides less modularity for the previous snapshots but has relatively a good result for the last snapshots t_f.

Table 1. The ANR-Info-RSN data sets characteristics

Data sets	t_0	t_f	$|N_s|$	$|E_s|$	$|N_f|$	$|E_f|$
DS1	July 17, 2014	July 31,2014	10569	14121	801	997
DS2	August 3, 2014	August 15,2014	6162	8069	390	451
DS3	August 17, 2014	August 31,2014	10189	12263	424	508
DS4	September 3, 2014	September 30,2014	8224	10371	412	535

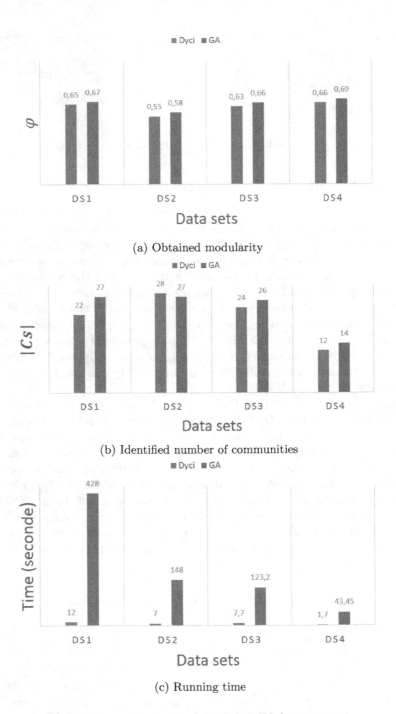

(a) Obtained modularity

(b) Identified number of communities

(c) Running time

Fig. 9. The results for the ANR-Info-RSN data sets at t_f

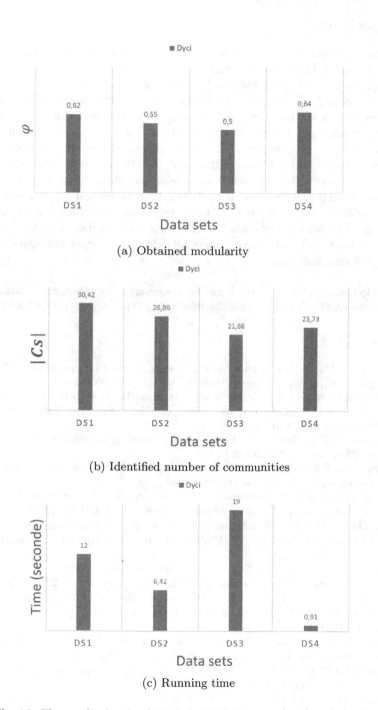

(a) Obtained modularity

(b) Identified number of communities

(c) Running time

Fig. 10. The results for the ANR-Info-RSN data sets for the whole Cs_s

6   Conclusion

To conclude, community identification in time-varying networks enhances our understanding of the graph structure over time. In this paper, a community detection algorithm for weighted and dynamic graphs, called *Dyci*, is proposed. The main idea of *Dyci* is to track whether a connected component of the weighted graph becomes weak (i.e., in terms of weight) over time, in order to merge it with the "dominant" neighbour community. In order to assess the quality of the identified communities by *Dyci*, a computational comparison is conducted with GA on real-world data sets of the ANR-Info-RSN project. The latter shows that *Dyci* obtains satisfying results with relatively short time. As perspectives, we project to extend this work to multi-graphs with several edges types linking two nodes. As an instance, considering retweet edges and mention edges at the same time. The latter exists when a Twitter's user mention another Twitter's user in its tweet. In this context, community detection algorithm could consider overlapping communities.

Acknowledgments. This research has been supported by the Agence Nationale de la Recherche (ANR, France) during the Info-RSN Project (ANR-13-SOIN-0008).

References

1. Harary, F., Gupta, G.: Dynamic graph models. Math. Comput. Model. **25**(7), 79–87 (1997). http://dx.doi.org/10.1016/S0895-7177(97)00050-2
2. Diehl, S., Görg, C.: Graphs, they are changing. In: Goodrich, M.T., Kobourov, S.G. (eds.) GD 2002. LNCS, vol. 2528, pp. 23–31. Springer, Heidelberg (2002). doi:10.1007/3-540-36151-0_3
3. Newman, M.E.J., Girvan, M.: Finding and evaluating community structure in networks. Phys. Rev. E **69**, 26113 (2004)
4. Blondel, V., Guillaume, J., Lambiotte, R., Mech, E.: Fast unfolding of communities in large networks. J. Stat. Mech. **10**, 10008 (2008)
5. Tantipathananandh, C., Berger-Wolf, T., Kempe, D.: A framework for community identification in dynamic social networks. In: KDD 2007: Proceedings of the 13th ACM SIGKDD International Conference on Knowledge Discovery and Data Mining, pp. 717–726. NY, USA (2007). http://portal.acm.org/citation.cfm?doid=1281192.1281269
6. Takaffoli, M., Sangi, F., Fagnan, J., Zane, O.R.: Community evolution mining in dynamic social networks. Procedia Soc. Behav. Sci. **22**, 49–58 (2011). dynamics of Social Networks 7th Conference on Applications of Social Network Analysis-ASNA2010. http://www.sciencedirect.com/science/article/pii/S1877042811013784
7. Bansal, S., Bhowmick, S., Paymal, P.: Fast community detection for dynamic complex networks. In: F. Costa, L., Evsukoff, A., Mangioni, G., Menezes, R. (eds.) CompleNet 2010. CCIS, vol. 116, pp. 196–207. Springer, Heidelberg (2011). doi:10.1007/978-3-642-25501-4_20
8. Aktunc, R., Toroslu, I.H., Ozer, M., Davulcu, H.: A dynamic modularity based community detection algorithm for large-scale networks: Dslm. In: Pei, J., Silvestri, F., Tang, J. (eds.) ASONAM, pp. 1177–1183. ACM (2015). http://dblp.uni-trier.de/db/conf/asunam/asonam2015.html#AktuncTOD15

9. Nguyen, N.P., Dinh, T.N., Xuan, Y., Thai, M.T.: Adaptive algorithms for detecting community structure in dynamic social networks. In: INFOCOM, pp. 2282–2290. IEEE (2011). http://dblp.uni-trier.de/db/conf/infocom/infocom2011.html#NguyenDXT11

10. Alvari, H., Hajibagheri, A., Sukthankar, G.R.: Community detection in dynamic social networks: A game-theoretic approach. In: Wu, X., Ester, M., Xu, G. (eds.) ASONAM, pp. 101–107. IEEE Computer Society (2014). http://dblp.uni-trier.de/db/conf/asunam/asonam2014.html#AlvariHS14

11. Tantipathananandh, C., Berger-Wolf, T.Y.: Constant-factor approximation algorithms for identifying dynamic communities. In: Iv, J.F.E., Fogelman-Souli, F., Flach, P.A., Zaki, M. (eds.) KDD, pp. 827–836. ACM (2009). http://dblp.uni-trier.de/db/conf/kdd/kdd2009.html#TantipathananandhB09

12. Hopcroft, J., Khan, O., Kulis, B., Selman, B.: Tracking evolving communities in large linked networks. In: PNAS (2004)

13. Greene, D., Doyle, D., Cunningham, P.: Tracking the evolution of communities in dynamic social networks. In: Proceedings of the 2010 International Conference on Advances in Social Networks Analysis and Mining, pp. 176–183. ASONAM 2010, (2010). http://dx.doi.org/10.1109/ASONAM.2010.17

14. Xie, J., Chen, M., Szymanski, B.K.: Labelrankt: Incremental community detection in dynamic networks via label propagation. CoRR abs/1305.2006 (2013). http://dblp.uni-trier.de/db/journals/corr/corr1305.html#abs-1305-2006

15. Abdelsadek, Y., Chelghoum, K., Herrmann, F., Kacem, I., Otjacques, B.: Community detection algorithm based on weighted maximum triangle packing. In: Proceedings of International Conference on Computer and Industrial Engineering CIE45 (2015)

16. Park, Y., Song, M.: A genetic algorithm for clustering problems. In: Koza, J.R., Banzhaf, W., Chellapilla, K., Deb, K., Dorigo, M., Fogel, D.B., Garzon, M.H., Goldberg, D.E., Iba, H., Riolo, R. (eds.) Genetic Programming 1998: Proceedings of the Third Annual Conference, pp. 568–575. Morgan Kaufmann, University of Wisconsin, Madison, Wisconsin, USA, 22–25 July 1998

17. Pizzuti, C.: GA-Net: a genetic algorithm for community detection in social networks. In: Rudolph, G., Jansen, T., Beume, N., Lucas, S., Poloni, C. (eds.) PPSN 2008. LNCS, vol. 5199, pp. 1081–1090. Springer, Heidelberg (2008). doi:10.1007/978-3-540-87700-4_107

18. Jin, D., He, D., Liu, D., Baquero, C.: Genetic algorithm with local search for community mining in complex networks. In: ICTAI (1), pp. 105–112. IEEE Computer Society (2010). http://dblp.uni-trier.de/db/conf/ictai/ictai2010-1.html#JinHLB10

19. Newman, M.: Modularity and community structure in networks. Proc. Nat. Acad. Sci. **103**(23), 8577–8582 (2006)

20. Abdelsadek, Y., Chelghoum, K., Herrmann, F., Kacem, I., Otjacques, B.: Visual interactive approach for mining twitter's networks. In: Tan, Y., Shi, Y. (eds.) Data Mining and Big Data. LNCS, vol. 9714, pp. 342–349. Springer, Heidelberg (2016)

Author Index

Printed in the United States
By Bookmasters

Printed in the United States
By Bookmasters